Practice Problems for the Environmental Fundamentals of Engineering Exam

240 Practice Problems with Solutions

John T. Fox, Ph.D., P.E.

Foreword

Studying environmental engineering is a great start to an incredibly rewarding career. Few career options afford the opportunity to perform work which has such a tremendous benefit to public health. Your career will also offer an opportunity to utilize an incredibly diverse range of academic topics. Environmental engineering requires the applied understanding of chemistry, physics, statistics, biology, public health, public policy, air quality, mathematics, aquatic chemistry, geology, toxicology, water treatment, organic chemistry, wastewater treatment, and environmental science (just to name a few). Due to the requirements for environmental engineers to study an incredibly broad range of topics, environmental engineering can be a very prescriptive major – yet still not cover every topic on the fundamentals of engineering exam during your undergraduate education.

The Fundamentals of Engineering Exam is an excellent way to advance your career. Over the years, many environmental engineering students have requested materials on various topics in order to help them prepare for the environmental engineering FE exam. This book is in response to those requests from students. Depending on your elective choices, you may not have selected electives that directly map to all of the FE exam topical areas. This set of practice problems has been developed based on the NCEES Environmental FE Exam Specifications.

Remember, you have been studying environmental engineering during your undergraduate studies and have been preparing for this exam for years. The review content and review questions will help you refresh and focus for the FE exam. Good luck!

About the Author

John Fox is currently an Assistant Professor at Lehigh University. At Lehigh University, John works in the Department of Civil and Environmental Engineering, which offers an ABET accredited B.S. in Environmental Engineering. At Lehigh University, John has taught seven different courses including: Hazardous Waste Treatment and Management (CEE 378), Fundamentals of Air Pollution (CEE 373), Environmental Water Chemistry (CEE 274), Environmental Engineering Design (CEE 377), Environmental, Geotechnics and Hydraulics Laboratory (CEE 275), Environmental Organic Chemistry (CEE 396), and Environmental Case Studies (CEE 379).

In addition to teaching, John is an active researcher pursuing research at the intersection of material science and environmental engineering. Areas of interest include; preventing VOC air pollution from manufacturers, advancing low-VOC adhesive technologies, diesel particulate filter materials, activated carbon, and recycling waste materials for beneficial reuse.

And just like you, in the spring of 2006 the author was an undergraduate engineering student who studied for the Fundamentals of Engineering Exam, and passed. Following graduate work and gaining the required experience the author passed the PE exam.

Education

Virginia Military Institute	B.S.	Civil & Environmental Engineering
The Pennsylvania State University	M.S.	Environmental Engineering
The Pennsylvania State University	Ph.D.	Environmental Engineering

Getting Started

First, take a deep breath, you purchased this book and are actively preparing for your FE exam. You are moving in the right direction.

Second, take a moment and create an account with NCEES (https://ncees.org/) and download the FE Supplied Reference Handbook from NCEES. Once you create your account, follow the link within your dashboard under the "Useful Documents" section click on "Reference Handbooks" and download the most recent FE Reference Handbook. Spend time familiarizing yourself with the FE Reference Handbook. The FE is a computer based test (CBT), so your reference handbook will also be available in a digital format. It is important that you are comfortable working with the digital format of the reference book, performing calculations by hand, and entering your answer via the computer.

Next, review the topics provided by NCEES that will be on your environmental FE:

https://ncees.org/wp-content/uploads/FE-Env-CBT-specs.pdf

Now, let's take a look at these topics carefully. You will notice that NCEES provides a range of the number of questions that fall under each topic. There are five topics that the author would consider to be general engineering topics, and these include; "Mathematics" (4-6), "Probability and Statistics" (3-5), "Ethics and Professional Practice" (5-8), "Engineering Economics" (4-6), and "Thermodynamics" (3-5). The topics comprising the traditional environmental engineering topical areas include; "Materials Science" (3-5), "Environmental Science and Chemistry" (11-17), "Risk Assessment" (5-8), "Fluid Mechanics" (9-14), "Water Resources" (10-15), "Water and Wastewater" (14-21), "Air Quality" (10-15), "Solid and Hazardous Waste" (10-15), and "Groundwater and Soils" (9-4). Within these nine environmental engineering topical areas, there could be as few as 81 questions or as many as 91, based on the minimum number range of questions in these core environmental engineering topical areas and the minimum number range of questions in the general engineering. With the FE exam containing 110 questions, 74-83% of the environmental engineering exam will focus on these core topical areas. Therefore, you should devote a majority of your time studying these core areas! As such, these practice problems will be focused predominately on these topics, with a few practice questions in the general engineering topical questions.

Tips for Solving Problems

When you are facing the FE, it is best to have a few useful tips on how to give yourself the best opportunity to arrive at a correct answer. First, take your time reading each problem. I spend countless hours every year explaining homework problems or test problems that students have made incorrect assumptions because they misread the problem. Take your time and make sure you know what you are answering. Now, once you know what you are answering, there are generally three broad approaches to FE problems. The approaches to answering problems fall under what I call, "knowledge", "dimensional analysis", and "equation based" problems. In the following paragraphs we will explore each of these approaches.

"Knowledge" questions are problems I classify as, you either know or don't know, or you must use logic to determine an appropriate answer. Examples of "knowledge" questions would be definitions of terms, or applying definitions to problems, or using your acquired knowledge to 'select the best answer'. These are generally questions where you do not have to perform any calculations or use any equations.

"Dimensional Analysis" questions are problems which do not require a formalized equation, per se, but require you to inventory data from the problem statement and manipulate the data into a useful value. These types of problems require you to utilize parameters of mass, time, energy, volume, consumption, etc. – in order to arrive at an answer. A simple example would be: A small town of 10,000 people produce 3 pounds of trash per capita per day, what is the mass of trash produced per week. In this type of problem, do not spend your valuable time searching for an equation! Simply recognize that this problem can be answered through dimensional analysis and perform the calculation with: *10,000 people x 3 lbs/person day x 7 days/wk = 210,000 lbs/week*. These problems challenge you to recognize the data provided and how to use relationships between the provided data in order to correctly manipulate that data through calculations to find the correct answer.

"Equation Based" questions are problems which require you to utilize a formalized equation from the appropriate discipline of the question being asked. In these types of problems you will be provided with questions that ask a very specific calculation to be performed which requires the selection of the correct equation, the correct input of values into the equation, and the successful calculation of an answer.

Also, you can have a type of question that utilizes any combination of the three types of approaches to solving problems. As long as you recognize what type of approach you should take for a problem, you have provided yourself with an improved likelihood to arrive at a correct answer.

Tips for Preparing for the FE

As you prepare for the FE it is best to recognize that this is a test which is very different than other tests you have taken as an undergraduate. In fact, you could think of this as the ultimate final exam that tests all of your knowledge acquired in math, science, and engineering courses. As an undergraduate you may have been able to study the night before the exam and perform well enough to pass your class or perhaps achieve an outstanding grade. The FE is not an exam you can cram for at the last minute. In fact, your preparation for the FE takes the entirety of your undergraduate study in order to best prepare you for the exam. Below are few common do's and don'ts to help your FE preparation.

- Don't wait until the night before to study.
- Do develop a study plan by topical area and spend several weeks reviewing key topics.

- Don't only focus your review on topical areas which you have had courses.
- Do study topical areas that you have not yet had courses.

- Don't assume that the availability of the "Reference Handbook" for your test will assure that you will select the correct equation.
- Do spend ample time familiarizing yourself with the "Reference Handbook" and understand the location and format of equations.

- Don't spend most of your time reviewing topics that are a small percentage of the FE.
- Do spend most of your time reviewing topics that are a majority of the FE questions.

- Don't stay up all night before the FE pulling an all-nighter.
- Do go to bed early and get a restful night of sleep. Plus, eat a healthy breakfast!

- Don't arrive on time to your testing center.
- Do arrive early to the testing center as this will help you stay relaxed. It is very stressful to be in traffic when you are already supposed to be somewhere.

- Don't buy a calculator for the FE exam and only use it for the exam.
- Do buy a calculator for the FE exam (or use your current calculator if it is approved) and perform practice problem calculations using the same calculator you will use for the FE.

Table of Contents

Section Name	Page Number
Section 1. Mathematics	1
Section 2. Probability and Statistics	3
Section 3. Ethics and Professional Practice	5
Section 4. Engineering Economics	9
Section 5. Materials Science	10
Section 6. Environmental Science and Chemistry	12
Section 7. Risk Assessment	17
Section 8. Fluid Mechanics	19
Section 9. Thermodynamics	24
Section 10. Water Resources	26
Section 11. Water and Wastewater	35
Section 12. Air Quality	49
Section 13. Solid and Hazardous Waste	57
Section 14. Groundwater and Soils	66
Answer Key	71

1- Mathematics

1.1 What is the surface area of the solid box with a cylindrical hole, as depicted below?

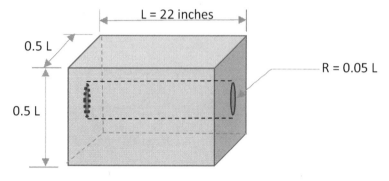

a. 968 in²
b. 1202 in²
c. 1354 in²
d. 1538 in²

1.2 Find the derivative of $y = x^5 + 12x^3 + 81x$

a. $\frac{dy}{dx} = 5x^5 + 36x^3 + 81x$
b. $\frac{dy}{dx} = 5x^4 + 36x^2 + 81$
c. $\frac{dy}{dx} = 5x^5 + 144x^2 + 81x$
d. $\frac{dy}{dx} = 20x^4 + 72x^2 + 81$

1.3 Find the derivative of $y = \frac{9x^3}{5} + 5x^2 + 8x$

a. $\frac{dy}{dx} = \frac{27x^2}{5} + 10x^2 + 8x$
b. $\frac{dy}{dx} = \frac{27x^3}{5} + 10x^2 + 8x$
c. $\frac{dy}{dx} = \frac{27x^2}{5} + 10x + 8$
d. $\frac{dy}{dx} = \frac{27x^3}{5} + 10x^2 + 8$

1.4 Find the roots of the following equation $x^2 - 10x + 16 = 0$

    a. (x-4) and (x+4)
    b. (x-8) and (x-2)
    c. (x-10) and (x-6)
    d. (x+4) and (x-14)

1.5 What is the difference of the two matrices provided below

$$\begin{bmatrix} 85 & 37 \\ 22 & 14 \end{bmatrix} - \begin{bmatrix} 54 & -14 \\ 8 & 2 \end{bmatrix} =$$

    a. $\begin{bmatrix} 31 & 51 \\ 14 & 12 \end{bmatrix}$
    b. $\begin{bmatrix} 31 & 23 \\ 30 & 16 \end{bmatrix}$
    c. $\begin{bmatrix} 54 & -14 \\ 8 & 2 \end{bmatrix}$
    d. $\begin{bmatrix} -31 & -51 \\ -14 & 12 \end{bmatrix}$

1.6 What is the product of the two matrices provided below

$$\begin{bmatrix} 11 & 4 \\ 8 & 3 \end{bmatrix} \cdot \begin{bmatrix} 2 & 7 \\ 9 & 22 \end{bmatrix} =$$

    a. $\begin{bmatrix} 22 & 28 \\ 72 & 66 \end{bmatrix}$
    b. $\begin{bmatrix} 36 & 77 \\ 16 & 56 \end{bmatrix}$
    c. $\begin{bmatrix} 22 & 78 \\ 27 & 66 \end{bmatrix}$
    d. $\begin{bmatrix} 58 & 165 \\ 43 & 122 \end{bmatrix}$

2 - Probability and Statistics

2.1 You have a large bucket that contains equally sized 240 red cubes and 60 green cubes. The lid does not allow you to see the cubes, but you can reach in with your hands to retrieve a cube. What is the probability that you reach in and retrieve only 2 cubes of the same color on your first attempt?

    a. 3.94%
    b. 50.0%
    c. 67.9%
    d. 95.7%

2.2 You are performing experiments on photo-degradation of organic compounds in water. The compounds are exposed to light for 24 hours and then each reactor is immediately sampled three times. You have 10 reactors running for these experiments. What is your sample size for this data set?

    a. 10
    b. 13
    c. 30
    d. unable to tell from the information provided

2.3 In the following data set of 12 numbers, which value best represents the mode?

40	45
40	45
40	45
40	50
40	50
45	50

    a. 40
    b. 44
    c. 45
    d. 50

2.4 In the following data set of 13 numbers, which value best represents the median?

11, 15, 16, 16, 19, 22, 23, 28, 31, 35, 36, 38, 41

a. 16
b. 23
c. 25
d. 28

2.5 In the following data set of numbers, what is the mean?

18, 19, 35, 36, 38, 40, 40, 41, 42, 42, 42, 49, 49, 51, 55, 63, 71

a. 40
b. 41
c. 42
d. 43

2.6 You are analyzing experimental data. The value of α is established as 0.05 prior to starting the experiment. After reviewing the data you find a p-value to be 0.03. Based on your analysis what can you say about the statistical significance of the experimental data?

a. The data is not statistically significant.
b. The value of α is too low, and should never be set at this level.
c. The α value and p-value are not related. Therefore there is not enough context in this problem.
d. The data is statistically significant.

2.7 You are analyzing lead in drinking water in the homes of a community. The sample you just tested has a value of 20 ppb of lead. The average in the community is 16 ppb with a standard deviation of 2 ppb. Based on your analysis what percentage of homes in the community have a higher value of lead, than 20 ppb?

a. 0.13%
b. 2.28%
c. 21.2%
d. 18.4%

3 – Engineering Ethics

3.1 Your firm is quickly growing and needs to hire someone new that will mostly perform field work and construction oversite. Your boss asks you to convince your friend from college to come join your firm. Your boss confirms that your friend will work in the field, but tells you that you can receive a $2500 bonus if your friend joins your firm. When talking with your friend, your friend tells you that they want to mostly work on design projects. What statement from the Engineering Code of Ethics best describes why it is wrong to tell your friend that they would mostly be performing design work in order to entice them to join your firm?

    a. Engineers shall disclose all known or potential conflicts of interest that could influence or appear to influence their judgment or the quality of their services.

    b. Engineers shall not solicit or accept financial or other valuable consideration, directly or indirectly, from outside agents in connection with the work for which they are responsible.

    c. Engineers shall not reveal facts, data, or information without the prior consent of the client or employer except as authorized or required by law or this Code.

    d. Engineers shall not attempt to attract an engineer from another employer by false or misleading pretenses.

3.2. You recently changed jobs to a new engineering firm. Your old engineering firm wants to continue to pay you on a limited basis to answer about 1 hour per week of phone calls from a client who trusts you and needs assistance from you on a treatment system you helped design and construct. What do you think is the best and most appropriate way to handle this situation?

    a. Continue to answer the phone calls and not tell your new engineering firm, as it is not a client of your new firm, and you know they would fully understand that the client trusts you and needs your assistance on the treatment system.

    b. Ignore the phone calls, because you don't think it is right to be paid by two different engineering firms at the same time.

    c. Fully disclose the situation to your new employer and with their approval continue to help the client who trusts you.

    d. Stop answering emails and phone calls from your old engineering firm asking you to help with this client, since you are no longer working there.

3.3. A large privately held local business uses your competing firm for all of their engineering work. Can you file a complaint that the work is not available for bid?

a. Yes, engineering work should be available for public bidding.

b. No, clients are not legally required to seek bids for engineering services.

c. Yes, because it is the best interest of the profession.

d. No, because you hope your own firm would receive the work, and that would be promoting your own best interest.

3.4. You have the opportunity for a very large client to employ your firm for engineering services as they expand their business in a high growth area. This client is known for paying very well and working for this client is viewed very favorably. You will be one of the leading engineers working for this client in this new business area. Your boss has put tremendous pressure on you, for the importance of landing this prospective client. The prospective client has confided in you that there are only two firms they are seeking to employ. Which of the following statements would be an appropriate way to comment on this information that the prospective client has shared with you?

a. Discuss the key outcomes of your prior assignments and how these past accomplishments align with the goal of this new business area.

b. Discuss with the prospective client how you believe the other firm being considered does not perform work as well as your firm.

c. Discuss "off the record" with your contact from the prospective client about possible opportunities to share bonus money you receive if they award the work to your firm.

d. Discuss with the prospective client how you used to work for the other firm being considered by the prospective client, and compare and contrast the difference between the two firms.

3.5. Considering air quality regulations, what best describes an area that is in non-attainment with regard to NAAQS?

a. A region that exceeds the regulated concentrations for one or more of the seven criteria pollutants.

b. A region has one or more area sources that are not in compliance with their current permit.

c. A region which possesses one or more point locations that are emitting air pollution without a permit.

d. An area which has increased ambient concentrations of carbon dioxide.

3.6. What best defines how maximum contaminant level goals (MCLG) are established as non-enforceable public health goals in drinking water.

a. The highest level of a level of a contaminant in drinking water at which no known or anticipated adverse effect on the health of persons would occur.

b. This is the highest level allowed of disinfectant or disinfection by-product allowed in public water, other compounds are not considered governed by MCLGs.

c. The MCLG is set by using a specified treatment technique that will achieve 99.9% removal of a specified compound, and this is how all MCLs are established.

d. The MCLG of a compound is established based on public consensus.

3.7. The management of hazardous waste that is currently being generated is regulated at a federal level by which law:

a. CERCLA – Comprehensive Environmental Response, Compensation, and Liability Act

b. Both the Montreal Protocol and Kyoto Protocol.

c. RCRA – Resource Conversation and Recovery Act

d. The US Superfund Program

3.8. To what end does professional engineering licensing serve the best interest of the U.S. public?

a. As the title professional engineer implies, only those with licensing which includes education and experience are qualified to practice engineering that impacts public safety, public welfare, safety, safeguarding of life, health.

b. It helps keep the cost of engineering low and therefore keeps taxes low.

c. It ensures that all engineers are fully capable, based on their degrees to work as engineers so that public projects are completed in a timely manner.

d. Profession Engineering licensing ensures that those that achieve this status are part of a small professional community that are connected to other engineers through public work.

3.9. What type of permit is required for any point source that discharges water with "pollutants" into a water body of the United States?

    a. United States Clean Water Act Permit (US CWA Permit)

    b. Maximum Contaminant Level (MCL)

    c. Resource Conversation and Recovery Act (RCRA)

    d. National Pollutant Discharge Elimination System (NPDES)

4 – Engineering Economics

4.1 If Carson deposits $10,000 in an account that pays 4.0% compounded annually. How many years will it take for his initial deposit to double?

    a. 10 years
    b. 18 years
    c. 22 years
    d. 72 years

4.2. A piece of specialty analytical equipment that your firm is looking to purchase costs $8,500. The piece of equipment has a 10 year useful life, but requires $750 per year of maintenance, calibration, and service. After 10 years, the equipment can be salvaged for $350. If interest rates are 6.0%, what is the present worth of the costs associated with this analytical equipment?

    a. $15,653
    b. $13,827
    c. $12,753
    d. $11,657

4.3. Your company is looking to enter the filter market. You have 525 customers who will purchase 1 of your filters per month. The filter requires tooling and other equipment that necessitates a $135,000 up-front capital cost. After that capital expense, each filter costs your firm $15 to make, but you can sell the filters for $29.50. In what month does your company break-even?

    a. month 1
    b. month 9
    c. month 18
    d. month 27

4.4. Calculate the incremental costs per filter for production given the following scenario. A filter company produces 100,000 filters at a total cost of production of $324,000. If the filter company plans to produce 120,000 filters, the total production cost is $346,000. What is the incremental cost per filter for the additional 20,000 filters?

    a. $0.18
    b. $1.10
    c. $2.88
    d. $17.30

5 – Materials Science

5.1 Why might iron be specified instead of aluminum in an application that experiences temperatures between 400 and 800°C?

    a. Iron is stronger in tension and is less likely to warp than aluminum at high temperatures.
    b. Iron is easier to machine
    c. Iron is more economic per pound, and both are adequate at high temperatures.
    d. Iron has a higher melt point than aluminum

5.2 Which chemical would not likely cause accelerated corrosion of iron?

    a. Water
    b. Oxygen
    c. Sodium chloride
    d. Epoxy resin

5.3 What is the classification of materials that possess the electrical conductivity between that of a metal and that of an insulator?

    a. Super conductors
    b. Strain insulators
    c. Transformers
    d. Semiconductors

5.4 What physical material property of activated carbon best predicts adsorption performance for an organic compound that has low water solubility?

    a. Specific density
    b. Specific heat
    c. Surface area
    d. Luster

5.5 What is the engineering strain in a steel rod that measures 2.0000 inches that elongates to 2.0021 inches?

    a. 0.00001
    b. 0.00105
    c. 0.0021
    d. 1.00105

5.6 What is the mass of 2 cubic meters of silver?

   a. 2,000 kg
   b. 5,698 kg
   c. 17,838 kg
   d. 21,000 kg

5.7 What is the resistance of a resistor which uses an aluminum wire in the resistor that possesses a cross sectional area of 1.5×10^{-5} m^2 and is 1.25 m long?

   a. 0.002 Ohms
   b. 0.004 Ohms
   c. 0.008 Ohms
   d. 0.010 Ohms

5.8 How much energy is required to raise the temperature of a 5 kg block of iron from 10°C to 300°C?

   a. 561 kJ
   b. 662 kJ
   c. 748 kJ
   d. 833 kJ

5.9 Which of the following is not a common method to corrosion control?
   a. Coatings and linings
   b. Cathodic Protection
   c. Material Selection
   d. Electromagnetic Induction

5.10 Which term describes when a material transitions from elastic to plastic behavior?

   a. Yield
   b. Stress
   c. Strain
   d. Ductile

6 – Environmental Science and Chemistry

6.1 A large warehouse has 3,200 µg/m³ of benzene (C_6H_6) in the air inside the warehouse is 25°C. If the K_H of benzene is 1.8×10^{-1} [M/atm] what is the expected concentration in an open 5 gallon bucket containing tap water that sits inside the warehouse?

   a. 0.014 mg/L
   b. 0.140 mg/L
   b. 1.40 mg/L
   d. 14.0 mg/L

6.2 What is the correct stoichiometric reaction for the complete oxidation of glucose ($C_6H_{12}O_6$)?

   a. $C_6H_{12}O_6 \rightarrow 3O_2 + 6H_2 + 6C$
   b. $C_6H_{12}O_6 \rightarrow 6CO_2 + 6H_2O$
   c. $C_6H_{12}O_6 + O_2 \rightarrow 6CO_2 + 6H_2O$
   d. $C_6H_{12}O_6 + 6O_2 \rightarrow 6CO_2 + 6H_2O$

6.3 Balance the following redox reaction.

   FeS + HCl → FeCl₂ + H₂S

   a. 2 FeS + 3 HCl → 2 FeCl₂ + H₂S
   b. FeS + 2 HCl → FeCl₂ + H₂S
   c. FeS + 2 HCl → FeCl₂ + 2 H₂S
   d. 2 FeS + HCl → 2 FeCl₂ + H₂S

6.4 Balance the following chemical reaction.

   FeCl₃ + H₂O → Fe(OH)₃ + HCl

   a. 2 FeCl₃ + 3 H₂O → 2 Fe(OH)₃ + 3 HCl
   b. 3 FeCl₃ + 2 H₂O → 2 Fe(OH)₃ + 3 HCl
   c. 5 FeCl₃ + 2 H₂O → 3 Fe(OH)₃ + 5 HCl
   d. FeCl₃ + 3 H₂O → Fe(OH)₃ + 3 HCl

6.5 Cadmium is present at 4.7 mg/L in water. At what pH will the cadmium precipitate? The Ksp and precipitation reaction is provided as:

$$K_{sp} = 5.27 \times 10^{-15}, \quad Cd(OH)_2 \rightarrow Cd^{2+} + 2\ OH^-$$

a. 1.12
b. 4.95
c. 9.05
d. 12.88

6.6 Acetic acid has a pKa of 4.74. If a solution contains 0.5 M acetate and 0.5 M acetic acid, what is the pH in a closed system?

a. 3.74
b. 4.69
c. 4.74
d. 4.79

6.7 What is the pH of a solution that contains 0.003 M H^+ in a closed system.

a. 1.51
b. 1.80
c. 2.52
d. 3.74

6.8 What is the numeric stoichiometric ratio of oxygen to fuel (oxygen: fuel) in the complete combustion of propane (C_3H_8)?

a. 1:5
b. 1:1
c. 3:1
d. 5:1

6.9 The degradation of phenol in an aqueous system is determined to have the rate constant (k) of 0.055 (1/hours). The initial concentration of phenol in the aqueous system is 130 mg/L. When will the concentration of phenol be 65 mg/L?

a. 3.575 hours
b. 7.150 hours
c. 12.60 hours
d. 1180 hours

6.10 In a wastewater treatment batch process, NH_4-N is converted to NO_3^- at a constant rate as depicted in the data below. Based on this data, what is the rate of NH_4-N conversion to NO_3^-?

NH_4-N	NO_3^-	Time
24 ppm	14 ppm	1.5 hours
14 ppm	26 ppm	3 hours

a. 4 ppm/hour
b. 8 ppm/hour
c. 12 ppm/hour
d. 14 ppm/hour

6.11 Acetone contains a carbonyl functional group. How does this carbonyl functional group influence the environmental characteristics of acetone?

a. Polar and less water soluble
b. Non-polar and less water soluble
c. Polar and more water soluble
d. Non-polar and more water soluble

6.12 A benzene ring contains a methyl-functional group at the 1st and 3rd position, what is the correct nomenclature for this specific xylene?

a. ortho-xylene
b. para-xylene
c. meta-xylene
d. terra-xylene

6.13 Phytoextraction is best defined as:

a. A process by which metal contaminants are removed from soil via plants.
b. A process where plants stabilize soils with their root systems.
c. A process where plants release compounds from their roots.
d. A process where plants are extracted from the ground to promote bio extraction.

6.14 Eutrophication has been identified as degrading the water quality in a river. What contaminants should be prevented from entering the river, from both point and non-point sources, to stop eutrophication?

a. BOD and COD
b. BOD and TSS
c. Nitrogen and Phosphorous
d. Bacteria and Algae

6.15 The octonal water partitioning coefficient can be used to provide insight of organic chemicals and what environmental characteristics?

   a. Water solubility and volatility
   b. Ion Exchange behavior and water solubility
   c. Activated Carbon Adsorption and Partitioning from water to organic soils
   d. The water solubility of organics in the presence of ionic species.

6.16 A two liter liquid vessel contains 0.75 L water (s.g. 1.0) and 0.75 L olive oil (s.g. 0.85). In an experiment, 20 mg of a chemical labeled "C" was poured into this two liter liquid vessel. The Kow for "C" is 0.05. What is the concentration of compound "C" in the water?

   a. 26.7 mg/L
   b. 25.3 mg/L
   c. 1.27 mg/L
   d. 0.11 mg/L

6.17 The dimensionless Henry's solubility constant (Cw/Ca) for ethanol is 4,885. In a laboratory the organic compound is measured at 1 ppm, which is 1,885 micrograms of ethanol per cubic meter of air. What is the anticipated concentration (micrograms per liter) in an open beaker of water on the laboratory bench?

   a. 90.5 µg/L
   b. 905 µg/L
   c. 9,205 µg/L
   d. 19,205 µg/L

6.18 The Henry's solubility constant (Cw/Ca) for phenol is 39,000. In a room there is 3,850 micrograms of phenol per cubic meter. What is the anticipated concentration (mg per L) in an open glass of water on a table in this room?

   a. 15 mg/L
   b. 150 mg/L
   c. 185 mg/L
   d. 380 mg/L

6.19 Balance the following chemical reaction:

Fe + Cl$_2$ → FeCl$_3$

a. 3 Fe + 2 Cl$_2$ → 3 FeCl$_3$
b. 2 Fe + 3 Cl$_2$ → 2 FeCl$_3$
c. 4 Fe + 5 Cl$_2$ → 4 FeCl$_3$
d. 4 Fe + 3 Cl$_2$ → 4 FeCl$_3$

6.20 Balance the following chemical reaction:

KMnO$_4$ + HCl → KCl + MnCl$_2$ + H$_2$O + Cl$_2$

a. 2 KMnO$_4$ + 4 HCl → 3 KCl + 2 MnCl$_2$ + 4 H$_2$O + 5 Cl$_2$
b. 2 KMnO$_4$ + 5 HCl → 3 KCl + 2 MnCl$_2$ + 8 H$_2$O + Cl$_2$
c. 2 KMnO$_4$ + 8 HCl → KCl + 2 MnCl$_2$ + 4 H$_2$O + Cl$_2$
d. 2 KMnO$_4$ + 16 HCl → 2 KCl + 2 MnCl$_2$ + 8 H$_2$O + 5 Cl$_2$

6.21 The minimum dissolved oxygen level in rivers to support fish is generally accepted to be what range?

a. 0-2 mg/L
b. 2-4 mg/L
c. 4-6 mg/L
d. 8-10 mg/L

6.22 The dissolved oxygen deficit for a river is 3.2 mg/L, which was calculated from the Streeter Phelps equation. The saturated dissolved oxygen concentration in the river is 9.8 mg/L. What is the dissolved oxygen concentration in this river?

a. 13.0 mg/L
b. 9.8 mg/L
c. 6.6 mg/L
d. 3.2 mg/L

7 – Risk Assessment

7.1 Which of the following is not an exposure route considered for toxicology?

    a. auditory
    b. inhalation
    c. dermal
    d. ingestion

7.2 Chlorine water disinfection can produce chloroform as a disinfection by-product. Chloroform is reasonably anticipated to be a human carcinogen, and the Oral S.F. is 6.1×10^{-3} A female resident has lived in a town for 22 years with water that contains 40 µg/L of chloroform. What is her chronic daily intake from drinking this water?

    a. 4.4×10^{-2} mg/kg d
    b. 4.4×10^{-4} mg/kg d
    c. 4.1×10^{-6} mg/kg d
    d. 4.1×10^{-8} mg/kg d

7.3 Certain states require that employees of gas stations pump gas. What is the chronic daily intake of benzene for a male gas station service employee, who works five 8 hour days per week for 50 weeks per year. The employee starting working at the gas station at 18 and just turned 47. The average ambient concentration of benzene is 0.65 mg/m³. Benzene is classified as a human carcinogen.

    a. 0.001 mg/kg d
    b. 0.011 mg/kg d
    c. 0.112 mg/kg d
    d. 1.12 mg/kg d

7.4 When performing the feasibility study for site remediation, the EPA risk assessment guidelines set forth a preliminary remediation goals based on risk-based calculations, and chemical concentrations for carcinogens are equivalent to a lifetime cancer risk of what value?

    a. Zero
    b. 1×10^{-9}
    c. 1×10^{-6}
    d. 1 in 100

7.5 What is the carcinogenic risk associated with drinking water contaminated with benzene for an individual who has a quantified chronic daily intake of 5×10^{-4} mg/kg d. The Oral S.F. for benzene can be assumed to be 2.9×10^{-2} (kg day/mg).

  a. 1.86×10^{-2}
  b. 1 in 100
  b. 1.45×10^{-5}
  c. 53.7

7.6 Find the chronic daily intake for the consumption of drinking water for 15 years with 1.0 mg/L of acetone. The individual of interest is a 35 year old female. Acetone has a noncarcinogenic Oral RfD of 0.9 mg/kg d.

  a. 0.033 mg/kg d
  b. 0.029 mg/kg d
  c. 0.007 mg/kg d
  d. 0.0007 mg/kg d

7.7 When evaluating the hazard quotients for all chemicals for which an individual is exposed to, a total hazard index can be quantified. Which of the following total hazard index values represent a value at which no adverse human health effects would be expected to occur.

  a. 4.5
  b. 2.5
  c. 1.5
  d. 0.5

8 – Fluid Mechanics

8.1 A curved concrete ditch, which is a half-circle, has a radius of 36". The ditch has a slope of 4.75% and a roughness coefficient of 0.012. What is the maximum flow (cfm) that this ditch can carry before the ditch overflows the concrete sides?

    a. 5300 cfm
    b. 500 cfm
    c. 337 cfm
    d. 38 cfm

8.2 A 4" diameter pipe in a tall building will carry water to a holding tank 525 feet higher than the pump. A check valve will be installed after the pump, in order to prevent water flow back to the pump. What is the pressure exerted on the check valve due to the water in the pipe?

    a. 70 psi
    b. 228 psi
    c. 2250 psi
    d. 6430 psi

8.3 When specifying a type of pump, what is meant by the term "flooded suction"?

    a. The pump must pull water up from a depth below the pump in order to prime the pump.
    b. The pump is located at an elevation below the fluid source which flows into the suction side of the pump.
    c. The pump provides enough flow into a tank for overflow to occur, and this is when the tank is flooded.
    d. The pump is located so that when in operation there is no possibility for suction to occur.

8.4 What is the head loss in 1000' of 12" plastic pipe carrying 3200 gallons per minute?

    a. 17 ft
    b. 25 ft
    c. 39 ft
    d. 57 ft

8.5 A large water tank is at capacity, and holds 100,000 gallons. The tank has a radius of 9 feet. The tank has several 1 inch diameter outlets 4.5 feet from the ground, which have a coefficient of discharge of 0.65. These outlets can be used to attach devices for various applications. If one of the outlets is not plugged, and the effluent water freely flows into the atmosphere, what is the initial flow rate out of the full tank?

   a. 0.20 cfs
   b. 0.79 cfs
   c. 0.83 cfs
   d. 0.91 cfs

8.6 What is the flow rate for the submerged outlet described below?

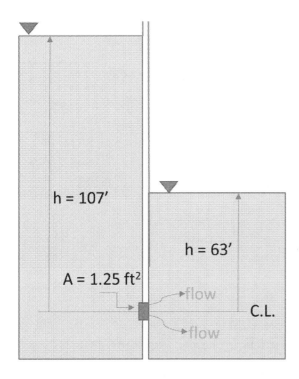

   a. 17.1 cfs
   b. 25.2 cfs
   c. 39.9 cfs
   d. 57.8 cfs

8.7 A pump is needed for an air stripping tower. The air stripping tower is 10 m tall, and requires a water flow rate of 200 L/sec. The pump will be operating at 71% efficiency. What is the approximate brake horse power requirement for the water pump?

   a. 13.9 kN m/sec
   b. 19.6 kN m/sec
   c. 27.6 kN m/sec
   d. 67.5 kN m/sec

8.8 What is the cost to operate a 10 hp pump for a year, when electricity is 5 cents per kwhr?

   a. $3300
   b. $2900
   c. $2500
   d. $2100

8.9 A municipality is going to install a pump station to lift 1.5 MGD of water 121 feet to a water storage tank. What is the pump horsepower requirement for a 74% efficient pump?

   a. 13 hp
   b. 23 hp
   c. 33 hp
   d. 43 hp

8.10   A municipality is going to install a pump station to lift 35 MGD of water 357 feet to a water storage tank. What is the pump horsepower requirement for a 77% efficient pump?

   a. 2900 hp
   b. 1800 hp
   c. 1400 hp
   d. 1035 hp

8.11   When designing for relatively low flows, what design practice is best for optimum pump efficiency?

   a. Design the biggest pump possible, it is best to oversize a pump than to undersize a pump
   b. Always pick the cheapest pump
   c. Use variable speed drives and smaller pumps to improve efficiency
   d. Find the pump that operates at the highest rpm, the faster a pump spins the better the water flows.

8.12   A 90° V-notch weir has 9.5 inches of head depth. What is the discharge in cfs?

   a. 0.78 cfs
   b. 1.42 cfs
   c. 390 cfs
   d. 707 cfs

8.13   A rectangular free discharge suppressed weir has a length of 1.75 feet. The head depth is 14.25 inches. What is the discharge in cfs?

   a. 1980 cfs
   b. 314 cfs
   c. 7.54 cfs
   d. 4.17 cfs

8.14   What is the flowrate from a fire-hydrant discharging into the atmosphere with an outlet diameter of 3 inches, a pressure of 25 psi, and a smooth and well-rounded outlet?

   a. 570 gpm
   b. 890 gpm
   c. 1120 gpm
   d. 1210 gpm

8.15 Two similar centrifugal pumps are operating in series. The pumps both have a 12" impeller and are operating at the same speed. Pump A places 367 feet of head on the system and Pump B places 349 feet of head in the series. What is the resultant head on the system?

    a. 18 feet
    b. 349 feet
    c. 367 feet
    d. 716 feet

8.16 What is the head loss in 5 miles of 6" cast iron pipe (about 20 years old), carrying 600 gallons per minute?

    a. 392 ft
    b. 878 ft
    c. 1241 ft
    d. 1732 ft

9 – Thermodynamics

9.1 How much energy is required to heat 37.5 L of water from 17.5°C to 90°C?

   a. 32.5 kJ
   b. 11,364 kJ
   c. 49,050 kJ
   d. 369,403 kJ

9.2 A company produces a baked product that is baked in their industrial ovens. The product is currently baked from room temperature (20°C) to 190°C on stainless steel trays that weigh 2.2 pounds each. The company is evaluating using identically sized aluminum trays that weigh 0.8 pounds each. What is energy use difference for the aluminum tray per baked product?

Stainless Tray Properties: 7.7 g/cm³   0.5 J/g°C
Aluminum Tray Properties: 2.7 g/cm³   0.9 J/g°C

   a. 29,400 J
   b. 32,930 J
   c. 55,540 J
   d. 62,070 J

9.3 A completely empty 55-gallon rigid drum is sealed at -13°C and at sea level. The drum will be shipped and eventually stored in a 60°C warehouse at sea level. What is the air pressure in the steel drum at the warehouse?

   a. 10.2 psi
   b. 14.7 psi
   c. 18.8 psi
   d. 67.8 psi

9.4 A large 125,000 square foot warehouse has a ceiling height of 15 feet. The near sea-level location keeps the warehouse temperature near 20°C, year round. If the warehouse was completely empty, how many pounds of air would there be in the warehouse?

    a. 4990 lbs
    b. 9640 lbs
    c. 13,900 lbs
    d. 141,000 lbs

9.5 An ideal gas at 15°C is initially held in a 20-gallon rigid container at ambient pressure. How much work can be performed if the gas is completely pressed down to 0.75 gallons during an isothermal process?

    a. 254.5 J
    b. 7,862 J
    c. 32,357 J
    d. 47,891 J

9.6 What is the mixture molecular weight for a gas that contains 20% oxygen, 35% nitrogen, and 45% carbon dioxide?

    a. 28 g/mole
    b. 29 g/mole
    c. 33 g/mole
    d. 36 g/mole

9.7 What is the mixture molecular weight for a gas that contains 10% oxygen and 90% carbon dioxide?

    a. 33.8 g/mole
    b. 39.7 g/mole
    c. 40.6 g/mole
    d. 42.8 g/mole

10 – Water Resources

10.1 A town of 25,000 people has a population which uses water based on household size. In this town it is reported that 10% of the population lives alone, 20% of the population lives in a 2 person household, 25% live in a household of 3 people, and 45% of the population lives in a household of 4 or more people. What is the daily water demand (MGD) for the town, based on the per capita use in the table below?

Household Size	Per capita water use
1 person	360 L/d
2 people	285 L/d
3 people	250 L/d
4 or more people	200 L/d

a. 1.05 MGD
b. 1.62 MGD
c. 2.81 MGD
d. 6.70 MGD

10.2 A town with a population of 11,200 use an average of 350 L/d per capita. This town has plans to build 3 new hotels and a new self-service laundry. In total, the three hotels will employ 36 employees and host 550 guests per night. The new self-service laundry will have a total of 20 washing machines. Based on the usage rates provided in the table, what percent of current daily water use will be added to the daily water demand?

Source	Unit	Flow rate, gal/unit day
Hotel	Guest	80
Hotel	Employee	15
Self-Service Laundry	Washing machine	425

a. 1.35%
b. 2.74%
c. 5.12%
d. 7.41%

10.3   A town has 15,000 residents, plus commercial water use that is currently 1,350 m³/day and is expected to remain constant. What is the calculated average and peak water demand for the town?

Use	Demand	Peaking Factor
Residential	90 gal/capita - day	2.5
Commercial	1,350 m³/day	1.7

   a. Average: 1.71 MGD, Peak: 3.98 MGD
   b. Average: 1.71 MGD, Peak: 5.67 MGD
   c. Average: 2.70 MGD, Peak: 3.98 MGD
   d. Average: 2.70 MGD, Peak: 5.67 MGD

10.4   A small town can be considered to provide only residential water. This town has currently maxed out their water allocation at 2.0 MGD. There are 20,000 people in the town who use an average of 100 gallons/capita –day. The town is considering implementing a water conservation program where the town will provide water-conserving shower heads. The water-conserving shower heads save use 20% less water for showering. The town expects 75% of the population to utilize the water saving shower heads. Based on the residential water use for this town, provided in the table, how much total water will be conserved per day with this water-conserving shower head program?

Residential Use	Percent of Total Water Use
Shower	18%
Faucet	17%
Dishwashing	2.5%
Clothes Washing	19%
Toilet Flushing	26%
Other Use	17.5%
Total	**100%**

   a. 51,840 gallons
   b. 54,000 gallons
   c. 57,600 gallons
   d. 72,000 gallons

10.5 A city has an average water demand of 35 MGD. The city water use is 82% residential and 18% commercial. The peaking factor for residential use is 2.9 and the peaking factor for commercial water use is 1.35. What is the daily peak flow rate in the city?

    a. 47.3 MGD
    b. 61.0 MGD
    c. 91.7 MGD
    d. 101.5 MGD

10.6 When precipitation falls, the land area that contributes to surface runoff to a specific watercourse or water body, is defined as what?

    a. Sheet flow
    b. watershed
    c. topographical area
    d. detention

10.7 When rain falls on a watershed such that the soil and vegetation uptake is exceeded, what is the term used to describe the water which exceeds uptake?

    a. Surface runoff
    b. Sheet flow
    c. Stream flow
    d. Hydrological flow

10.8 A 10-year rainfall event produces a 5.1 inches per hour, during a 30 minute duration, which is equal to the time of concentration. The 12.5 acre drainage area is currently a grassy field with a runoff coefficient of 0.33. If the drainage area is partially developed and the net runoff coefficient for the drainage area becomes 0.42. What is the increase in peak runoff rate for a 10-year storm, following development?

    a. 5.7 cfs
    b. 7.9 cfs
    c. 11.4 cfs
    d. 15.3 cfs

Use Figure 10.1, 10.2, and Table 10.1, 10.2 for questions 10.9-10.14.

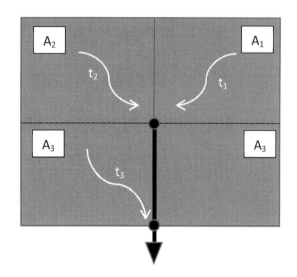

Figure 10.1. Hypothetical Drainage system.

Table 10.1. Drainage System Information

Area 1	Area 2	Area 3
11 Acres	12 Acres	24 Acres
t_1 = 17 min	t_2 = 22 min	t_3 = 28 min

Table 10.2. Runoff Coefficients

Land Use	Runoff Coefficient
Asphalt Street	0.80
Concrete Street	0.85
Roofs	0.82
Heavy Industrial	0.75
Light Industrial	0.61
Downtown Business	0.77
Single Family Residential	0.44
Suburban Residential	0.32
Playgrounds	0.31
Parks	0.16
Unimproved ground	0.12

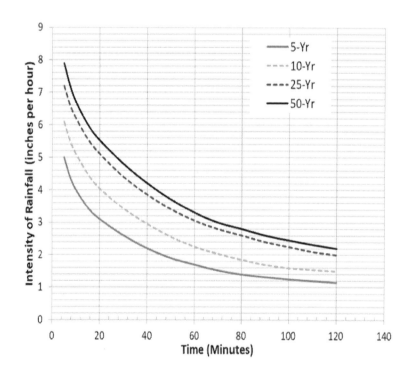

Figure 10.2. Intensity-Duration-Frequency curve for Drainage System illustrated in Figure 8.1.

10.9 Prior to any development in the drainage system depicted in Figure 8.1, what is the peak run-off for the system for a 50-year rainfall event?

    a. 9.87 cfs
    b. 18.0 cfs
    c. 20.3 cfs
    d. 28.2 cfs

10.10 A developer wants to turn Area 1 into a suburban residential area, Area 2 into a light industrial area, and the zoning board is requiring that Area 3 becomes a park. What is the peak run-off expected for a 25-year rainfall event?

    a. 507 cfs
    b. 143 cfs
    c. 79.5 cfs
    d. 49.9 cfs

10.11 If Area 2 becomes a heavy industrial zone, what is the expected peak flow leaving Area 2, for a 10-year rainfall event?

    a. 34.7 cfs
    b. 24.3 cfs
    c. 17.9 cfs
    d. 13.7 cfs

10.12 Area 1 is under review for rezoning. If Area 1 is used for single family residential land use, what is the peak runoff possible from Area 1 for a 50-year rainfall event?

    a. 21.3 cfs
    b. 28.1 cfs
    c. 37.4 cfs
    d. 45.2 cfs

10.13 A proposed development plan leaves Area 1 and Area 2 unimproved, and turns Area 3 into a light industrial complex. What is the maximum runoff from the drainage system for a 50-year rainfall event?

    a. 54.2 cfs
    b. 62.6 cfs
    c. 76.2 cfs
    d. 87.2 cfs

10.14   If all of Area 1 were to be paved with asphalt. What is the peak flow expected from Area 1 for a 5-year rainfall event?

   a. 13,000 gpm
   b. 15,000 gpm
   c. 18,000 gpm
   d. 21,000 gpm

10.15   A city uses 92 million gallons per day. The available withdrawal for the river that the city utilizes as a water source is provided, such that the available withdrawal provides a small safety factor between water allocation and natural flow. Based on the chart provided, which depicts a once in 20-year drought, what is the closest value of the required storage volume for the reservoir.

   a. 0.54×10^9 ft^3
   b. 1.54×10^9 ft^3
   c. 3.56×10^9 ft^3
   d. 9.62×10^{10} ft^3

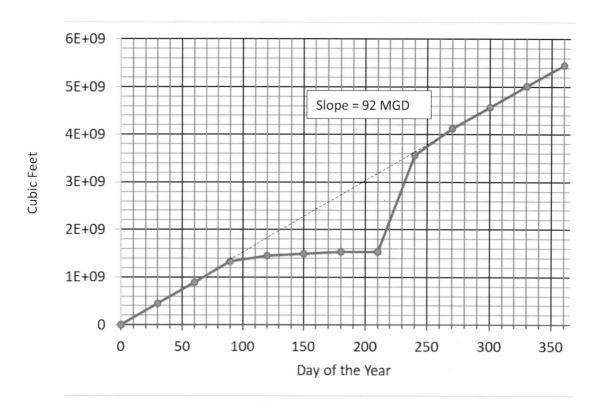

10.16 A town that utilizes wells as the water source use constant velocity pumps. Collectively, the wells produce a constant 750 gpm. The demand for the town is provided in Table A. What is the minimum operating volume for a reservoir to service this town?

Table A. Hourly demand for town described in problem 10.16.

Hour Ending	Demand (gpm)
0100	645
0200	602
0300	574
0400	541
0500	529
0600	608
0700	619
0800	637
0900	659
1000	708
1100	765
1200	821
1300	949
1400	1007
1500	1258
1600	1355
1700	1022
1800	742
1900	709
2000	688
2100	672
2200	658
2300	647
2400	679

a. 95,000 gal
b. 116,000 gal
c. 134,000 gal
d. 149,000 gal

10.17 Often, reservoirs are sized to provide enough capacity for a drought that occurs once in twenty years. What may happen if the 30-year rainfall data and river flow data utilized to develop reservoir sizing, happens to be the wettest 30-year period in a 500 year period?

   a. The reservoir may be oversized
   b. The reservoir will be sized correctly
   c. The reservoir will likely be undersized
   d. There is not enough information in the problem.

10.18 Water type of soil is most likely to erode due to a rainfall or runoff event?

   a. Sandy and Silty Soil
   b. High Content Clay Soil
   c. High organic matter soil with some clay
   d. Well drained and well-graded gravel with no silt

10.19 Vegetation cover helps to prevent erosion from runoff and rainfall events. Which of the following is not an erosion control role of vegetation cover?

   a. Keeping soil in place
   b. Slowing water runoff velocity
   c. Protecting soil from the direct impact of falling rain
   d. Removing nutrients from the soil

10.20 What type of pollutant will likely increase in water during a channel erosion event?

   a. Nutrients
   b. Oil and Grease
   c. Heavy Metals
   d. Sediment

10.21   Floatables are one type of pollutant generated by urban land use. Commonly, where do floatables originate?

   a. Litter and Improperly disposed refuse
   b. Gas stations and refueling stations
   c. Landscaping processes
   d. Agricultural chemical application and Livestock yards

10.22   Best management practices for storm water quality management typically fall under two categories. What are these two categories?

   a. Non-structural Controls and Structural Controls
   b. Gas station sources and non-gas station sources
   c. Total Loading and Non-Total Loading Controls
   d. Agricultural Controls and Non-Agricultural Controls

10.23   Which of the following would not be considered a pollution prevention strategy for storm water quality management?

   a. Use vegetation and ground cover for natural filtration of runoff
   b. Enclose or cover pollution-causing activities
   c. Minimize land clearing and excavation
   d. Be sure to disrupt steep slopes and areas with erodible soils

11 Water and Wastewater

11.1 What is an appropriate level of residual chlorine in potable drinking water?

   a. 0.5 mg/L
   b. 5 mg/L
   c. 50 mg/L
   d. 500 mg/L

11.2 What would be an expected temperature of water arriving to a wastewater treatment plant located in North Carolina in the summer?

   a. 2 °C
   b. 5 °C
   c. 18 °C
   d. 52 °C

11.3 What would be an expected TDS concentration in seawater?

   a. 1000 mg/L
   b. 3,500 mg/L
   c. 35,000 mg/L
   d. 100,000 mg/L

11.4 What concentration of BOD_5 would be typical for domestic wastewater?

   a. 25 mg/L
   b. 250 mg/L
   c. 2,500 mg/L
   d. 25,000 mg/L

11.5 What would be a typical ratio of nitrogen (as N) to phosphorous (as P) in domestic wastewater?

   a. 400 mg/L as N to 100 mg/L as P
   b. 100 mg/L as N to 400 mg/L as P
   c. 10 mg/L as N to 40 mg/L as P
   d. 40 mg/L as N to 10 mg/L as P

11.6 A municipal wastewater treatment plant currently receives 2 MGD of wastewater with an average BOD concentration Of 325 mg/L. If a large industrial facility is constructed, it will provide an additional 750,000 gallons per day of wastewater with an expected BOD concentration of 550 mg/L. What is the total expected daily load of BOD, from both existing wastewater sources and the new Industrial facility, to the wastewater treatment plant in pounds per day?

   a. 1580 pounds
   b. 4850 pounds
   c. 6320 pounds
   d. 8870 pounds

11.7 A wastewater treatment facility produces 20,000 pounds of sludge per day, as dry solids. The sludge is currently 12% solids. If the plant installs a dewatering system, that raises the solids content to 18% solids. What is the approximate daily flowrate of water produced from the sludge dewatering?

   a. 8500 gallons
   b. 6700 gallons
   c. 4900 gallons
   d. 3600 gallons

11.8 A drinking water treatment plant (WTP) utilizes two sources of water, Source A (a surface water source) and Source B (a groundwater source). The groundwater source can provide up to 500,000 gallons per day and the surface water can provide up to 750,000 gallons per day. Powdered activated carbon is used to remove natural organic matter (NOM). The activated carbon removes 40 mg of NOM per gram of activated carbon. The WTP receives a typical concentration of 22 mg/L of NOM from the surface water and 0.5 mg/L of NOM from the groundwater. The WTP must dose activated carbon to remove the NOM to prevent disinfection by-products. If the WTP needs 1 MGD of water, what is the least amount of activated carbon (pounds per day) that the WTP can utilize on a daily basis?

   a. 2350 pounds per day
   b. 3130 pounds per day
   c. 4320 pounds per day
   d. 5150 pounds per day

11.9 A water treatment plant lifts 200 gpm to a height of 15 feet to their headworks. If the WTP builds a new headworks that is 20 feet of lift for the pump, what is the increase in energy use for the new pump compared to the current pump?

a. 25%
b. 33%
c. 48%
d. 66%

11.10 A water treatment plant will use ferric coagulation. The WTP treats 1.25 MGD and the ferric flocs settle at 0.2 inches/minute. What is the required clarifier surface area for this WTP?

a. 3,230 sq. ft.
b. 5,620 sq. ft.
c. 6,980 sq. ft.
d. 8,550 sq. ft.

11.11 A waste water treatment plant will use primary clarification to suspended solids. The WWTP treats 2.75 MGD and the clarifier overflow rate is 700 gpd/ft^2. What is the expected suspended solids removal rate?

a. 52%
b. 58%
c. 64%
d. 66%

11.12 A Which unit process is not used to provide disinfection?

a. Chlorination
b. Ozonation
c. Ultra-Violet Light
d. Nitrification

11.13 What is the common term used to describe the biological layer that accumulates on slow sand filters used in water treatment?

   a. Wasserfilter
   b. Schmutzdecke
   c. Bakterienschicht
   d. Verschmutzung

11.14 A uniform sized sand media filtration bed is 1.5 meters deep. The diameter of the particles is 297 microns and the bed possesses 40% voids. The filter bed treats 0.5 MGD and the cross section of the bed is 3 m by 3 m. The drag coefficient of 48.5. What is the head loss through this filtration bed?

   a. 3.15 m
   b. 4.22 m
   c. 5.08 m
   d. 6.17 m

11.15 A slow sand filter is fluidized in order to backwash the filter bed. The backwash flow rate is 100 L/second. The cross-sectional area of the bed is 3 m x 3 m. The terminal settling velocity of the sand is 1 foot per second. What is the porosity of the bed while fluidized?

   a. 0.65
   b. 0.48
   c. 0.39
   d. 0.31

11.16 An industrial wastewater has a flowrate of 10,000 gallons per day. The wastewater contains 7 mg/L of styrene and must be treated to 0.1 ppm. An activated carbon possesses isotherm parameters of K = 120 and 1/n = 0.56 for styrene. If activated carbon was to be used to treat this wastewater, what is the daily carbon utilization rate?

   a. 12.3 pounds
   b. 15.4 pounds
   c. 17.4 pounds
   d. 23.1 pounds

11.17   What class of compounds is activated carbon best suited to remove?

a. nutrients
b. metals
c. particulates
d. trace organic

11.18   A water treatment system will utilize activated carbon as a unit process. The water treatment plant does not wish to change out the activated carbon more often than once every 60 days. If the carbon utilization rate is 320 pounds per day and the density of the activated carbon is 25 pounds per cubic foot. What volume vessel should be utilized for a two vessel activated carbon system?

a. 1920 cf
b. 400 cf
c. 160 cf
d. 130 cf

11.19   An activated carbon possesses a K = 32 and 1/n = 0.78 for an organic compound "C". The influent concentration of "C" into the activated carbon bed is 25 mg/L and the effluent concentration is 2 mg/L. What is the ultimate loading of the organic compound for this activated carbon?

a. 55 mg/g
b. 80 mg/g
c. 93 mg/g
d. 105 mg/g

11.20   Why is granular activated carbon utilized instead of powdered activated carbon in large vessels for water treatment?

a. GAC has faster kinetics
b. PAC is too expensive
c. There is more available information for GAC than PAC
d. PAC causes too much headloss across the vessel, compared to GAC

11.21 **b. 4.2**

11.22 **d. 7.5 m**

11.23 **a. 11**

11.24 **d. The air stripper performance would be more effective**

11.25 Consider an air stripper which needs improved performance. You only have the option of installing a (higher flow rate) larger pump or larger air compressor. Which do you choose to improve the performance of the air stripper?

    a. Neither, both options do not influence performance
    b. The air compressor, because this would increase the stripping factor
    c. The water pump, because this would increase the loading rate
    d. The water pump, because this would increase the Qw

11.26 A dry sludge (0% moisture) contains 0.5 mg/kg of zinc, and the sludge is to be land applied in an agricultural application. If the dewatered sludge (70% moisture) is to be deposited at a rate of 50 tons per acre per year, how much zinc is applied per year per acre?

    a. 6.8 grams
    b. 13,650 grams
    c. 27,220 grams
    d. 64,840 grams

11.27 A municipal WWTP currently produces a sludge that contains 16% solids. The WWTP currently produces 23 tons per day of sludge. It costs the municipality $30/ton for landfill disposal of each ton of sludge. The WWTP can increase solids content via sludge de-watering to 21% solids, but will continue to perform landfill disposal. What is the annual savings in landfill tipping fees attributable to sludge-dewatering?

    a. $23,000
    b. $49,000
    c. $60,000
    d. $74,000

11.28 A municipality which can withdraw 5.5 MGD in total from all wells servicing the water authority. Population growth is driving water demand near the maximum capacity of the wells. The municipality is evaluating various water conservation approaches. Approximately 80% of the residences in the water authority have water metering. Option 1 is to install water meters, as users save an average of 30% in daily water use when customers have water meters. Option 2 is to provide a new garden hose nozzles to 100% of homes, which reduce water use for lawn watering by 15% for all homes. Option 3 is to provide water shower heads to 65% of homes which will reduce shower water use by 50%. Option 4 is to incentivize residences to replace older toilets, which is expected to reduce toilet flushing water by 50% in the 30% of homes expected to partake in the incentive.

Activity	Daily Water Use (gallons per residence)
Toilet Flushing	35
Shower	28
Faucets	25
Clothe Washer	24
Lawn/Garden Care	8
Bathing	3
Dishwasher	1.9

a. Option 1
b. Option 2
c. Option 3
d. Option 4

11.29 A municipality has increasing water demand and this demand is nearing the maximum capacity of the water supply. Water use in the municipality is 95% residential. The municipality is evaluating various water conservation approaches. Which of the following approaches would not likely reduce water use in residences?

a. Utilizing a tiered water billing rate, where high users pay more.
b. Utilizing water metering in all residences.
c. Providing a tax rebate to offset the purchase of residential water saving technologies, such as high efficiency appliances or automatic faucets.
d. Utilizing a flat billing rate structure for users, where all users pay the same for water regardless of use.

11.30 A wastewater treatment plant will use flocculation as a unit process. The flocculator tank is 7500 L and the wastewater temperature is 30°C. If the tank is provided with 50 W, what is the velocity gradient for this process?

    a. 31 s^{-1}
    b. 57 s^{-1}
    c. 74 s^{-1}
    d. 91 s^{-1}

11.31 What is the hardness (as mg/L of $CaCO_3$) of water with the following composition?

Cation	Concentration mg/L	Anion	Concentration mg/L
Na^+	49	Cl^-	48
Ca^{2+}	38	SO_4^{2-}	12
Mg^{2+}	6.1	Alkalinity	30

    a. 366
    b. 183
    c. 153
    d. 120

11.32 A water contains 105 mg/L of hardness as $CaCO_3$. If the water is softened by slaked lime, $(Ca(OH)_2)$. How much slaked lime must be added if the lime is 96% pure.

    a. 105 mg/L
    b. 80 mg/L
    c. 65 mg/L
    d. 51 mg/L

11.33 A completely mixed activated sludge process measured 2,850 MLSS, mg/L. The aeration basin is 350 m^3 and the water temperature is 21°C. The tank removes the waste sludge and effluent sludge at the same location providing a waste sludge flowrate of 12,500 L/d and the effluent flowrate of 37,500 L/d. What is the solids retention time in this completely mixed activated sludge process?

    a. 11 days
    b. 7 days
    c. 5 days
    d. 3 days

11.34 A conventional activated sludge process utilizes 55 m³ of air per kg of BOD₅. The flowrate in the activated sludge process is 1.5 MGD. The solids residence time is 5.7 days and the yield coefficient is 0.72. The mixed liquor suspended solids have been measured at 1,675 mg/L. What is the solids loading rate?

   a. 6600 L/min g
   a. 5400 L/min g
   b. 4100 L/min g
   c. 2900 L/min g

11.35 A tapered aeration process is plug-flow, with a mean cell residence time of 8.2 days. The volume of the tank is 2300 cubic meters and the flow rate is 575 m³/hour and the recycle ratio is 0.48. The yield coefficient is 0.9 kg biomass/kg BOD, the microbial death ratio is 0.05 day⁻¹ and the influent BOD concentration is 0.31 kg BOD₅ per m³ and the BOD₅ removal efficiency is 88%. What is the expected mixed liquor suspended solids concentration?

   a. 3100 mg/L
   b. 2010 mg/L
   c. 1730 mg/L
   d. 1220 mg/L

11.36 A community of 250 residences is planned to be built 7 miles from an existing centralized wastewater infrastructure. The cost to install piping from the community to the centralized wastewater tie-in is $1.5M per mile. One alternative is to have each residence construct an on-site sand mound which requires a capital cost of $25,000 per residence. Another alternative is to utilize a decentralized treatment system that can handle up to 100,000 gallons per day is proposed for $1.75M. Which following outcomes is likely for this new community?

   a. Tie into the centralized wastewater infrastructure.
   b. Construct an on-site sand mound for each residence.
   c. Install the decentralized treatment system to serve the community.
   d. None of the options provided adequately protect the environment.

11.37   A fixed film biotower treatment system treats municipal wastewater. The municipal wastewater arrives to the biotower with a BOD5 of 295 mg/L. The treatability constant for the wastewater is 0.06 min-1 and the wastewater is not recycled to the biotower. The biotower depth is 30 feet. The flow rate through the biotower is 100 ft³/min and the cross sectional area of the biotower is 75 ft². The coefficient for the modular plastic media is 0.5. What is the effluent BOD5 concentration?

   a. 246 mg/L
   b. 183 mg/L
   c. 109 mg/L
   d. 39 mg/L

11.38   A disinfection tank utilizes free chlorine in a baffled inlet with some intra-basin baffles. The hydraulic residence time in the disinfection tank is 15 minutes. The residual disinfection concentration at peak hourly flow is 0.35 mg/L. What is the inactivation for bacteria in this unit process?

	Unit	1-log	2-log	3-log	4-log
Protozoa	mg min/L	20-30	35-45	70-80	
Bacteria	mg min/L	0.1-0.2	0.4-0.8	1.5-3	10-12

   a. 1 -log
   b. 2 - log
   c. 3 - log
   d. 4 - log

11.39   A disinfection tank utilizes free chlorine in a baffled inlet with a perforated inlet baffle and perforated intra-basin baffles. The residual disinfection concentration at peak hourly flow is 0.55 mg/L. The water pH is 7.8 and the water temperature is 15°C. What is the minimum residence time in order to meet the required log reduction for viruses?

   a. 36 minutes
   b. 27 minutes
   c. 18 minutes
   d. 9 minutes

11.40 Which is the most appropriate order of unit processes for a municipal wastewater treatment plant?

   a. Fine screens, course screens, equalization, grit removal, primary treatment
   b. Primary treatment, equalization, grit removal, fine screens, course screens
   c. Equalization, fine screens, course screens, primary treatment, grit removal
   d. Course screens, fine screens, grit removal, equalization, primary treatment

11.41 Which is the most appropriate order of unit processes for a municipal wastewater treatment plant?

   a. Primary Settling, Biological Processes, Secondary Settling, Filtration, Disinfection
   b. Filtration, Disinfection, Biological Processes, Primary Settling, Secondary Settling
   c. Primary settling, disinfection, Biological Processes, Filtration, Secondary Settling
   d. Primary settling, Secondary Settling, Biological Processes, Filtration, Disinfection

11.42 An aerated lagoon that is will treat a wastewater with 590 mg/L of BOD, with a daily flowrate of 15,000 L/day. The overall first-order BOD removal rate(k) is 0.1 d⁻¹. If the effluent BOD concentration must be less than 20 mg/L, what is the minimum lagoon size?

The lagoon can be modeled with the equation.

$$\frac{S_e}{S_o} = \frac{1}{1+k\tau}$$

Where Se and So = the effluent (Se) and influent (So) BOD concentration, g/m³
k = the overall first-order BOD removal rate constant, d⁻¹
τ = the hydraulic retention time (V/Q), d

   a. 4275 m³
   b. 3450 m³
   c. 2625 m³
   d. 1800 m³

11.43 A city is located on a large lake. The volume of the lake has been quantified as 175,000 m3. The daily withdrawal rate from the lake at one end for water use is 16,500 m³. The wastewater plant discharges at the other end of the lake at 15,750 m³ per day. The lake is fed by a spring which makes-up the difference for evaporation and city consumption, which keeps the lake at a near constant volume of water. How many days after water is discharged into the lake as wastewater does it become source water for the city?

    a. 156 days
    b. 87 days
    c. 34 days
    d. 11 days

11.44 A thin membrane separates two solutions, the one solution is a brackish water. The brackish water contains 4,000 mg/L of TDS. The coefficient of water permeation is 3.1×10^{-5} [kmol/(m² s Pa)]. The net operating pressure ($\Delta P_a - \Delta \Pi$) is 1620 kPa. What is the water flux across the thin membrane?

    a. 25.1 kmol/m² sec
    b. 34.9 kmol/m² sec
    c. 50.1 kmol/m² sec
    d. 71.6 kmol/m² sec

11.45 Nitrification is a two-step biological oxidation process, which the stoichiometry is provided below. Based on this reaction, what is the theoretical oxygen demand (pounds per day) for complete denitrification for a wastewater flowing at 2.5 MGD and contains 30 mg/L of NH_4^+?

$NH_4^+ + 2O_2 \rightarrow NO_3^- + 2H^+ + H_2O$

    a. 2225 lbs of oxygen/day
    b. 1750 lbs of oxygen/day
    c. 1375 lbs of oxygen/day
    d. 1050 lbs of oxygen/day

11.46 Nitrifying bacteria are sensitive, as both *Nitrobacter* and *Nitrosomonas* are sensitive to wide range of environmental factors. Which environmental factor is most likely to strongly inhibit nitrification processes?

    a. A shift in operating pH from 7.0 to 7.5
    b. The introduction of 0.5 mg/L of chromium and 2.0 mg/L of copper.
    c. An increase in temperature from 18°C to 23°C.
    d. An increase would of 2.5 mg/L in dissolved oxygen concentrations.

11.47 For a cation exchange resin, which order represents the typical series for selectivity?

    a. $H^+ < Zn^{2+} < K^+ < Na^+$
    b. $Zn^{2+} < K^+ < Na^+ < H^+$
    c. $H^+ < Na^+ < K^+ < Zn^{2+}$
    d. $Na^+ < K^+ < Zn^{2+} < H^+$

11.48 A water containing 26 mg/L of zinc is to be treated by ion exchange. The water daily water flowrate is 12,000 L/day and the exchange capacity of the resin for zinc (Zn^{2+}, MW = 65.g/mole) is 375 meq/kg. What is the required volume of resin, if the resin has a specific gravity of 0.6?

    a. 125 L
    b. 104 L
    c. 72 L
    d. 43 L

11.49 A water containing 22 mg/L of an organic compound. The activated carbon column is 1.5 meters deep, the total volume of water treated at exhaustion is 20,450 L. The total volume treated at breakthrough is 17,835 L. What is the depth of the sorption zone?

    a. 1.30 m
    b. 1.04 m
    c. 0.72 m
    d. 0.20 m

11.50 A water containing 120 mg/L of benzene is to be treated by activated carbon. The benzene isotherm parameters are K = 0.05 and a = 155. If the effluent water must be treated to 5 mg/L. What is mass of carbon utilized per 100 liters of water?

    a. 125 g
    b. 371 g
    c. 723 g
    d. 937 g

12 - Air Quality

12.1 A specialty manufacturing company is evaluating purchasing an ESP. The company will need an ESP able to treat 100,000 acfm at 95% efficiency. The effective drift velocity will be 0.5 ft/sec, calculate the required plate area in ft².

   a. 599 ft²
   b. 9985 ft²
   c. 59,914 ft²
   d. 599,146 ft²

12.2 Determine the sulfur dioxide emission index (EI_{SO2} – g SO2/g Fuel) for a coal with an approximate chemical analysis of ($C_{55}H_{62}S$ – MW 755 g/mole). You may assume that all sulfur is completely combusted to form sulfur dioxide.

   a. 0.008 g SO_2/g Fuel
   b. 0.042 g SO_2/g Fuel
   c. 0.085 g SO_2/g Fuel
   d. 0.164 g SO_2/g Fuel

12.3 A commercial painting facility needs to determine their total VOC reduction. The facility currently operates with a capture efficiency of 70.0%. The captured air is then run through an incinerator with 99.0% destruction. What is the total VOC reduction?

   a. 69.3%
   b. 84.5%
   c. 99.0%
   d. 169%

12.4 An incinerator possesses an influent of one principle organic hazardous contaminant, consisting of 91 lbs/hr of Benzene. The effluent possesses 0.91 kg/hr of Benzene. What is the destruction and removal efficiency of the incinerator?

   a. 99.99%
   b. 97.80%
   c. 90.09%
   d. 88.99%

12.5 Given the choice between a cyclone process or baghouse, please indicate which process you would choose for:

Scenario 1: A low volume air flow, with large dust particles, necessitating moderate efficiency.
Scenario 2: A High volume air flow, with mostly very small dust particles, necessitating high efficiency.

a. Baghouse for Scenario 1 and Cyclone for Scenario 2
b. Baghouse for both Scenario 1 and 2.
c. Cyclones for both Scenario 1 and 2.
d. Cyclone for Scenario 1 and Baghouse for Scenario 2.

12.6 **Which best describes the key difference between a primary pollutant and a secondary pollutant.**
a. A primary pollutant is emitted directly from a source whereas a secondary pollutant is formed via atmospheric reactions.
b. A primary pollutant is derived from materials and a secondary pollutant is formed during combustion.
c. A primary pollutant is emitted in higher volumes from factories, whereas a secondary pollutant is emitted in low volumes from factories.
d. A primary pollutant is from stationary sources, whereas a secondary pollutant is emitted from mobile sources.

12.7 Determine the minimum cloth area for a reverse-air baghouse to filter 40,000 m³/min of air with 50 g/m³ of flour.

a. 8,000 m²
b. 36,000 m²
c. 44,444 m²
d. 98,323 m²

12.8 An elevated storage tank that holds acetone has been ruptured due to a construction accident at the facility, and is now leaking 5,087 g/sec of acetone on a continuous basis. The tank is 25 m off the ground and the plume rises an additional 10 m above the tank. There is a town that is 1.0 km downwind of the stationary tank (which is acting as a point source). The following atmospheric conditions are provided: wind velocity is 6.0 m/sec; stability class D. What is the value that closest represents the ground level concentration of acetone at the town?

   a. 30 mg/m³
   b. 62 mg/m³
   c. 938 mg/m³
   d. 1875 mg/m³

12.9 The working capacity of an activated carbon is 17 g benzene/100 g carbon. The carbon has a bulk density of 30 lb/cubic foot. What volume of carbon is necessary to adsorb 100 pounds of benzene?

   a. 6 ft³
   b. 20 ft³
   c. 38 ft³
   d. 590 ft³

12.10 An industrial facility has an emissions stack with an effective height of 50 m which emits 300 g/sec of xylene on a continuous basis. There is a town that is 2.0 km downwind of the facility. The following atmospheric conditions are provided: wind velocity is 9.0 m/sec; stability class C. What is the value that closest represents the ground level concentration of xylene at the town?

   a. 125 µg/m³
   b. 135 µg/m³
   c. 260 µg/m³
   d. 345 µg/m³

12.11 An electric generation facility combusts pure methane (CH_4). The air used for combustion is dry and contains 3.76 moles of nitrogen for each mole of oxygen. Assuming perfect stoichiometric combustion, what is the mole fraction of CO_2 in the wet exhaust stream?

a. 0.045
b. 0.065
c. 0.095
d. 0.333

12.12 A propane grill company needs to learn more about the water vapor from combustion in their grills. The grills use pure propane (C_3H_8). Assuming the air used for combustion is dry and contains 3.76 moles of nitrogen for each mole of oxygen. Assuming perfect stoichiometric combustion, what is the mole fraction of water in the wet exhaust stream?

a. 0.058
b. 0.116
c. 0.155
d. 0.571

12.13 What is the difference between thermal NOx and fuel NOx?

a. Thermal NOx is formed in the atmosphere and fuel NOx is from combustion
b. Thermal NOx is more stable at high temperatures than fuel NOx
c. Thermal NOx is formed from no combustion reactions
d. Thermal NOx is formed at high combustion temperatures and fuel NOx is from combustion with nitrogen bound in the fuel.

12.14 An activated carbon has the ultimate loading of 16 lbs of Benzene per 100 pounds of GAC. If the carbon is sued to treat an exhaust stream with 400 ppm of Benzene and a flowrate of 20 m³ per min, 1 atm of pressure, and a temperature of 25°C. How much carbon is used in 8 hours?

a. 170 lbs
b. 105 lbs
c. 85 lbs
d. 60 lbs

12.15   Activated carbon is best suited for removing which type air pollutants?

a. Nitrogen and Sulfur Oxides
b. Particulates
c. Polar Volatile Organics
d. Non-Polar Volatile Organics

12.16   A GAC treatment system will remove ethyl-benzene. The system is designed to remove 50 lbs of ethyl-benzene, from the exhaust stream between each GAC regeneration. If the GAC is 30 lbs/ft3 and can load 11 pounds of ethyl-benzene per 100 pounds of GAC, what is the volume of the carbon required?

a. 15.15 ft^3
b. 19.19 ft^3
c. 24.24 ft^3
d. 28.28 ft^3

12.17   What is the correct stoichiometric combustion reaction for butane (C_4H_{10})?

a. C_4H_{10} + 1.5 O_2 → CO_2 + H_2O
b. 3 C_4H_{10} + O_2 → 12 CO_2 + 10 H_2O
c. C_4H_{10} + 6.5 O_2 → 4 CO_2 + 5 H_2O
d. C_4H_{10} + 5 O_2 → CO_2 + 10 H_2O

12.18   An activated carbon has an ultimate loading of 20 lbs of Toluene per 100 lbs of GAC and 5 pounds of isopropanol per 100 pounds of GAC. If exhausted air hits the GAC bed with 100 ppm of toluene and 100 ppm of isopropanol, which will breakthrough first?

a. There is not enough information stated in this problem
b. It occurs simultaneously, because the concentration is equal for both compounds
c. Toluene
d. Isopropanol

12.19 Indoor air pollution is often worse than outdoor air quality. If indoor air pollution, specifically VOCs, were to be remediated what technology is best?

  a. HEPA air filter
  b. Full Building De-Humidifier/Humidifier
  c. Fabric Filter
  d. Activated Carbon

12.20 What is the concentration (mg/m³) for 525 ppm of acetone (C_3H_6O)?

  a. 525 mg/m³
  b. 729 mg/m³
  c. 932 mg/m³
  d. 1244 mg/m³

12.21 What is the concentration (µg/m³) for 942 ppb of isopropanol (C_3H_8O)?

  a. 1870
  b. 2310
  c. 2960
  d. 3380

12.22 On a clear night with minimal cloudiness, there is a wind speed of 5.3 m/s, which is measured 10 m above the ground. What is the atmospheric stability?

  a. A
  b. B
  c. C
  d. D

12.23 What is radon gas?

  a. A fuel type gas
  b. A common radioactive noble gas
  c. A gas from combustion
  d. A very rare, but very safe gas

12.24 Why is it important to remediate radon gas from homes?

a. Radon has a very pungent odor, and must be removed for aesthetic reasons
b. Radon is highly combustible and can cause explosions
c. Radon has a high affinity for hemoglobin in the human body
d. Radon is the second leading cause of lung cancer

12.25 A small sedan will emit NOx at a rate of 0.2 g/mile. If a small city has 1,600 sedans that commute an average of 9 miles (round-trip) per day, how much NOx is emitted from sedans in this city, per day?

a. 2880 g
b. 3900 g
c. 7990 g
d. 14,400 g

12.26 What air pollutants are the major contributors to acid rain?

a. Hydrochloric acids
b. Common VOCs
c. NOx and SO_2
d. Ozone depleting compounds like (CFCs)

12.27 A combustion process emits 85,000 ppm of CO_2 and 3,200 ppm of CO. What is the combustion efficiency?

a. 3.76%
b. 85.0%
c. 88.2%
d. 96.4%

12.28 A certain sized cyclone is being considered by an industrial facility. The industrial facility must remove 80% of the U.S. Mesh 200 particles (74 microns). The cyclone removes 50% of particles that are U.S. Mesh 100 (149 microns). What percent of U.S. Mesh 200 particles will be removed by this cyclone?

a. 25%
b. 33%
c. 50%
d. 95%

12.29 A conventional cyclone is sized so that the cone length is 2.0 m. How many effective turns will the cyclone provide?

a. 2.75
b. 3.75
c. 5.50
d. 7.50

12.30 What is the cone length of a high efficiency cyclone that has an inlet width of 0.42 m?

a. 5.00 m
b. 3.50 m
c. 3.15 m
d. 2.75 m

13 – Solid Waste

13.1 A waste contains 10% moisture, 75% volatile matter, and 15% ash. The ultimate analysis is provided in the table, below. What value is closest to the HHV (BTU/lb) for the waste?

Ultimate Analysis	% weight
Moisture	10
Carbon	40
Hydrogen	8
Oxygen	22
Sulfur	5
Ash	15

a. 2500 BTU/Lb
b. 7500 BTU/Lb
c. 9700 BTU/Lb
d. 11,000 BTU/Lb

13.2 A small company produces a sludge which contains 82% moisture and 18% of a hazardous inorganic/metal blend. What stabilizing agent would be best in this scenario to solidify this waste?

a. Powdered activated carbon
b. A Proprietary Thermoplastic Polymer
c. Portland Cement
d. Asphalt

13.3 A town of 15,000 people generates 2 pounds of trash per person each day. The trash is collected once per week. The density of the non-compacted trash is 95 kilograms per cubic meter. What is the volume of trash discarded per week by the town?

a. 150 m³
b. 1000 m³
c. 2200 m³
d. 4900 m³

13.4   Which of the following items is not often a critical factor when evaluating site conditions for building a new landfill?

   a. Proximity to population Centers
   b. Public Acceptance
   c. Nearby Bridge Capacities
   d. Distance from a military base

13.5   Which environmental law requires generators to be responsible for the waste they produce from "Cradle to Grave"?

   a. Resource Conversation and Recovery Act
   b. Clean Water Act
   c. Comprehensive Environmental Response and Compensation Liability Act
   d. National Environmental Policy Act

13.6   What test method is used on wastes that are stabilized and solidified in order to determine if they are toxic?

   a. TCLP
   b. SEM
   c. UV-Spec
   d. GC-MS

13.7   A waste which contains a Principle Organic Hazardous Constituent (POHC) is incinerated. The POHC is present at 20% of the waste sludge that is incinerated. The waste sludge is fed to the incinerator at 63 pounds per hour. The concentration of the POHC in the exhaust gas is 0.025% by mass, and the exhaust gas at POHC measurement is 230°C, 1 atm, and has a mass flow rate of 4.7 pounds per minute. What is the DRE of the POHC?

   a. 99.99%
   b. 99.89%
   c. 99.44%
   d. 88.81%

13.8 A town generates 25,000 pounds per day of municipal solid waste that is sent to a landfill. The in-place density of the fill (refuse plus cover) is 1,200 lb/yd³. What volume is necessary (in cubic yards) for a 30 year operation period is most nearly what value?

a. 912,500
b. 525,255
c. 228,125
d. 182,500

13.9 Which of the following type of landfill requires a double liner?

a. Construction and Demolition Debris Landfill
b. Municipal Solid Waste Landfill
c. Hazardous Waste Secure Landfill
d. Recycling Facility Storage

13.10 A municipal solid waste landfill has a 55 cm clay liner. The clay liner is 0.25 porous and the clay is a bentonite base material. The leachate ranges from 37.5 cm higher than the clay liner. The coefficient of permeability of the clay material is 7.2×10^{-7} cm/second. Approximately, how many years until the leachate breaks-through the landfill clay liner?

a. 19 days
b. 131 days
c. 1,350 days
d. 19,500 days

13.11 A landfill possesses a cover which is 20 cm thick. The methane concentration in the landfill is 350,000 mg/m³. The porosity of the landfill liner is 20%. The diffusion coefficient for methane is 0.5 cm²/second. The concentration of methane at the surface of the landfill liner is 245,000 mg/m3. What is the next flux of methane and direction of flux?

a. 3.07×10^{-7} g/cm² second
b. 5.07×10^{-7} g/cm² second
c. 7.07×10^{-7} g/cm² second
d. 9.07×10^{-7} g/cm² second

13.12   Which of the following waste is hazardous?

   a. An organic waste with a flash point of 75°C
   b. An inorganic waste that has a pH of 11.9
   c. A waste that contains over 80% sodium bentonite
   d. A waste mixture that has a pH of 2.25, but contains 50% water

13.13   When is it required to utilize a "manifest" form?

   a. When shipping hazardous materials on the highway
   b. When a factory needs a new permit to produce hazardous materials
   c. When a hazardous material is inadvertently released as air pollution
   d. When shipping hazardous waste materials to disposal

13.14   A waste is placed in a landfill, and initially contains a specific weight of 1080 pounds per cubic yard. The overburden pressure applied during compaction equipment is 400 pounds per square inch. The empirical constants used in this facility and compaction equipment are 0.16 cubic yards per square inch and 0.0008 cubic yards per pound. What is the in place specific weight following the 400 psi compaction?

   a. 2120 pounds
   b. 1910 pounds
   c. 1480 pounds
   d. 1140 pounds

13.15   Why is clay often selected as the material for landfill liners?

   a. Clay possesses a very low coefficient of permeability
   b. Clay has a beneficial ion exchange capacity for leachate
   c. Clay has a high adsorption capacity for leachate contaminants
   d. Clay materials under high temperatures can form brick-like components

13.16　What is the approximate composition of gases in the effluent gas from a mature MSW landfill that is utilized in an energy production project?

    a. 90% Methane, 8% Carbon Dioxide, 2% Other Gases
    b. 80% Methane, 15% Carbon Dioxide, 5% Other Gases
    c. 70% Methane, 25% Carbon Dioxide, 5% Other Gases
    d. 50% Methane, 45% Carbon Dioxide, 5% Other Gases

13.17　Community Waste Disposal is often funded by the tax base, flat fees, or user fee. User fee is also known as Pay-As-You-Throw (PAYT), where residents pay for their waste based on what they throw away. Which of the following would you NOT expect for a community that is switching from tax base funded waste disposal to PAYT?

    a. An increase in total recycling rates
    b. A decrease in illegal dumping activities
    c. A reduction in average waste collected per year
    d. Less economic security for the municipality budget

13.18　An industrial company generates a solid waste that is a consistent quality product, that is uniform in chemical composition. What might be the most economically beneficial waste management strategy for the company?

    a. Perform additional on-site waste audits
    b. Change landfill vendors, to a less expensive vendor
    c. Perform material blending on –site to dispose of this waste simultaneously with other wastes
    d. Perform an on-site recycling program

13.19　What penalties are possible for an individual working for a company who falsifies records to suggest that a hazardous waste was disposed of as a non-hazardous waste?

    a. No penalty, the company is responsible
    b. Fines paid by the company
    c. Individual Fines and jail time
    d. No fines under RCRA, CERCLA is responsible for remediation

13.20   Beyond economic reasons, why is it important to recycle aluminum materials?

   a. Only recycled aluminum can create aerospace grade aluminum
   b. Recycled aluminum is tougher and needed for certain metallurgical applications
   c. Recycled aluminum is preferred for 25% of medical grade applications
   d. Recycled aluminum needs 90% less energy than new aluminum from bauxite ore

13.21   A waste sludge contains 8.3% by weight of benzene, the POHC. The sludge is produced at a rate of 1,000 gallons per day. The specific gravity of the waste is 0.925. How many kilograms of benzene are incinerated per day to meet the four-nine criteria?

   a. 3500
   b. 3150
   c. 323.98
   d. 291.16

13.22   With regard to CERCLA what is a PRP?

   a. Potential Remediation Property
   b. Potential Remediation Process
   c. Partially Remediated Property
   d. Potentially Responsible Party

13.23   What is the daily energy content of waste from a cafeteria that contains 500 pounds of food waste and 83 pounds of paper waste?

   a. 1.2×10^6 BTU
   b. 1.4×10^6 BTU
   c. 1.6×10^6 BTU
   d. 1.8×10^6 BTU

13.24 A municipal waste contains 40% food waste, 24% paper waste, 6% cardboard, 15% plastic, % wood, 7% glass, and 3% metal cans. What is the energy content per ton of waste?

   a. 9.1×10^6 BTU
   b. 10.9×10^6 BTU
   c. 13.7×10^6 BTU
   d. 15.4×10^6 BTU

13.25 A hazardous wastewater contains cadmium. Which chemical best removes cadmium from the water? You are provided with the solubility products for sodium sulfide (Na$_2$S), Sodium Hydroxide (NaOH), or sodium carbonate (Na$_2$CO$_3$) given the following:

$K_{sp} = 5.27 \times 10^{-15}$,   Cd(OH)$_2$ → Cd^{2+} + 2 OH$^-$
$K_{sp} = 6.18 \times 10^{-12}$,   CdCO$_3$ → Cd^{2+} + CO$_3^{2-}$
$K_{sp} = 1.40 \times 10^{-29}$,   CdS → Cd^{2+} + S^{2-}

   a. Not enough information provided
   b. Sodium Sulfide
   c. Sodium Hydroxide
   d. Sodium Carbonate

13.26 During the initial phase of site assessment it is necessary to utilize historical records for detailed information. Which question listed below is not likely obtainable or helpful for site assessment from historical records?

   a. What industrial activities have occurred at the Site
   b. What chemicals were used the Site
   c. Were there many disgruntled workers that performed illegal activities?
   d. Who were the property owners?

13.27 You are working at a facility that has many hazardous waste containers. One of the remediation contractors ask you if halogenated organics can be mixed with amines. What is the possible outcome?

   a. This is fine, there is no reaction expected.
   b. This is fine, the compounds will polymerize
   c. This is not acceptable, the compounds will generate heat and toxic gases.
   d. This is not acceptable, the compounds will explode.

13.28   Is it acceptable to mix an oxidizing acid with isocyanates?

   a. This is fine, there is no reaction expected.
   b. This is fine, the compounds will polymerize
   c. This is not acceptable, the compounds will solubilize toxic material.
   d. This is not acceptable, the compounds will generate heat, fire, and toxic gases.

13.29   Would a mixture of ethanol and ethylene with water?

   a. This is fine, there is no reaction expected.
   b. This is fine, the compounds will polymerize
   c. This is not acceptable, the compounds will solubilize toxic material.
   d. This is not acceptable, the compounds will generate heat, fire, and toxic gases.

13.30   What is the purpose of the National Priorities List?

   a. This is a list of worst sites identified for long-term cleanup.
   b. This is a list of compounds that are a national priority
   c. This is a list of companies that are prioritized for cleanup responsibility
   d. This is a list of outcomes, which are prioritized for environmental cleanup.

13.31   During site remediation process for CERCLA, what is the purpose of the RI/FS?

   a. The purpose of RI/FS is to determine who should pay for remediation.
   b. To assess the nature and extent of contamination, and understand treatability
   c. To determine if the site should be listed on the NPL.
   d. The purpose of the RI/FS is to remove the site from the NPL.

13.32   If there is a aqueous waste containing radioactive metals, what could be a first step, prior to long-term management?

   a. Precipitate and concentrate the radioactive metals.
   b. Nothing, radioactive wastes can only be managed.
   c. Properly mix the waste and water, then dispose to the sanitary sewer.
   d. Properly mix the waste and water, then send to a landfill.

13.33   Which of the following type of radioactive waste is most likely managed through a geological depository?

   a. High level wastes
   b. Mid level wastes
   c. Low level wastes
   d. Millings or 11(e)2 wastes

13.34   What is the purpose of geological depository for radioactive wastes?

   a. To provide a shallow hole for low level wastes.
   b. To provide a way to easily bury millings.
   c. To use natural occurring trenches for mid level disposal.
   d. To utilize deep underground isolation of dangerous radioactive wastes.

14 – Groundwater and Soils

14.1 During a site investigation an old leaking chemical waste tank was located. The chemical waste tank contained lindane and benzene. Which chemical likely has a plume which has traveled the furthest in the groundwater?

Property	Lindane	Benzene
Log Kow	3.90	2.12
MW (g/mole)	290.8	78.1
Solubility	7.80 mg/L	~1780 mg/L

a. They should be equal
b. Benzene
c. Lindane
d. There is not enough information provided

14.2 An old orchard site is about 3 acres in size. There is arsenic contamination in the first 8 inches of top soil. What is the best remediation alternative from the options below?

a. Soil Vapor Extraction
b. Pump and Treat
c. Bioremediation
d. Phytoremediation

14.3 The soil near an old chemical processing facility will be remediated. The contaminated soil contains mostly light-weight, volatile hydrocarbons. The depth to groundwater is 20 feet. Which remediation alternative is the best from the options below?

a. Soil Vapor Extraction
b. Pump and Treat
c. Bioremediation
d. Phytoremediation

14.4   The groundwater near an old fuel processing facility needs to be remediated. The contaminated groundwater contains the petroleum hydrocarbons, BTEX. Which remediation alternative is the best from the options below?

   a. Soil Vapor Extraction
   b. Pump and Treat
   c. Bioremediation
   d. Phytoremediation

14.5   An unconfined aquifer has a hydraulic conductivity of 0.007 ft/sec. The radius of the well is 6 inches. At a distance 38 feet from the center of the well the water is 104 feet above the bottom of the aquifer. At the well, the height of the water from the bottom of the of the aquifer is 67 feet. What is the flowrate of water drawn from the well?

   a. 13.7 cfs
   b. 32.1 cfs
   c. 48.3 cfs
   d. 59.6 cfs

14.6   A soil has a specific gravity of 2.65, and contains 5 cm³ of voids per 40 cm³ of soil. An 80 cm³ soil sample is holding 8 mL of water. What is the degree of saturation for this 80 cm³ sample?

   a. 10%
   b. 22.5%
   c. 62.5%
   d. 80%

14.7   A sample cylinder that is 20 cm long and a radius of 5 cm represents a soil composite, which will be tested for hydraulic conductivity. The hydraulic gradient is measured to be 19 and the flow rate of through the cylindrical sample is 0.45 cm³/hour. What is the hydraulic conductivity?

   a. 1.1×10^{-8} cm/sec
   b. 2.9×10^{-8} cm/sec
   c. 6.7×10^{-8} cm/sec
   d. 8.4×10^{-8} cm/sec

14.8   Which type of soil would you expect to have the lowest hydraulic conductivity?

a. Clean Gravel
b. Course Sand
c. Fine Sand
d. Clay

14.9   The groundwater table in a small town is 25 feet below the surface during the winter and is 35 feet below the surface during the summer. What is the thickness of the zone of intermittent saturation in this small town?

a. 10 feet
b. 25 feet
c. 35 feet
d. Not enough information in the problem statement

14.10   Calculate the hydraulic gradient for the following schematic, where point A is 1.55 miles from point B:

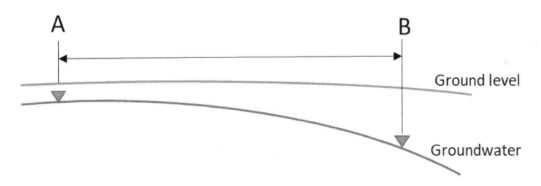

Point A

195 feet above MSL

GW is 5 feet below ground surface

Point B

191 feet above MSL

GW is 8 feet below ground surface

a. 0.0015
b. 0.0075
c. 0.00085
d. 0.00098

14.11 Which soil parameter will influence the water movement in an aquifer?

a. Cone of depression
b. Soil density
c. Soil Permeability
d. Soil Thickness

14.12 An artesian well with the potentiometric surface greater than the land surface refers to what type of scenario?

a. A carefully drilled well, named after Artisanal well drillers.
b. A well where water flows above the ground surface due to confined water pressure.
c. A well that pulls water from the surface into a confined aquifer.
d. A well that has the potential to utilize all water in a confined aquifer.

14.13 The hydraulic gradient for an aquifer is -0.0035. The hydraulic conductivity in the aquifer is 0.0058 cm/sec. The effective porosity in the aquifer is 0.175. What is the velocity of the groundwater in the aquifer?

a. 116 cm/yr
b. 366 cm/yr
c. 1.16 m/yr
d. 36.6 m/yr

14.14 The groundwater near a leaking underground storage tank is moving at 10.02 cm/day. A well is located 750 m from the underground storage tank. How long does it take for the contaminant to reach the well?

a. 75 days
b. 749 days
c. 7485 days
d. 74,847 days

14.15 A 1.35 kg soil sample is 0.5 L. The water content of the soil is 7.2%. What is the dry unit weight of the soil?

    a. 2.11 kg/L
    b. 2.31 kg/L
    c. 2.52 kg/L
    d. 2.79 kg/L

1-Mathematics

1.1	A_{box} = [22*(0.5*22) * 4] + {[(0.5*22)*(0.5*22) − (π(0.05*22)^2]*2} + (2π(0.05*22)*22) = 968 in^2 + 234.4 in^2 + 152 in^2 $\boxed{= 1354 \text{ in}^2}$
1.2	b. $\dfrac{dy}{dx} = 5x^4 + 36x^2 + 81$
1.3	c. $\dfrac{dy}{dx} = \dfrac{27x^2}{5} + 10x + 8$
1.4	b. (x-8) and (x-2)
1.5	a. $\begin{bmatrix} 31 & 51 \\ 14 & 12 \end{bmatrix}$
1.6	d. $\begin{bmatrix} 58 & 165 \\ 43 & 122 \end{bmatrix}$ (11*2) + (4*9) (11*7)+(4*22) (8*2) + (3*9) (8*7) + (3*22)

2 - Probability and Statistics

2.1	In this scenario, you need to consider both potential outcomes for drawing the same color cube. $$\left[\frac{240}{300} * \frac{239}{299}\right] + \left[\frac{60}{300} * \frac{59}{299}\right]$$ $$0.63946 + 0.03946$$ $$67.9\%$$
2.2	This is pure knowledge question. In this scenario, as described, there are 10 reactors and these represent your sample size for this data set. The fact that each reactor is sampled three times, is in fact a replication or repeated measurement, which does not change your sample size from the 10 reactors. The correct answer is a. 10
2.3	a. 40, this is the value that is repeated the most in the sample set.
2.4	b. 23, the median is the middle number in the list. (13+1)/2 = 7th number.
2.5	d. 43, the mean is the same as the average, which is the sum/n for any set of numbers.
2.6	d. The data is statistically significant. If the p-value is equal to or less than the ⍺, then the results are considered to be statistically significant.
27	a. 2.28%, For this answer you need to use Z = (X-μ)/σ = (20 −16)/2 = 2.0. Then, use the unit normal distribution table found in the FE Reference handbook, use 0.0228 from a Z of 2.0. 0.0228 = 2.28%.

3 – Engineering Ethics

3.1	d. Engineers shall not attempt to attract an engineer from another employer by false or misleading pretenses. *Answer: The correct answer is d, because if you are using false pretenses of "performing design work", when in fact most of the work will be in the field.*
3.2	c. Fully disclose the situation to your new employer and with their approval continue to help the client who trusts you. *Answer: The correct answer is c. It is best to always fully disclose all potential conflicts to your firm – which may include providing limited continuity on projects as you transition to a new firm. You should also make sure that your old firm and former client understand this is a temporary assistance, and they need to transition to be able to support this within their own capacity.*
3.3	b. No, private clients are not legally required to seek bids for engineering services.
3.4	a. Discuss the key outcomes of your prior assignments and how these past accomplishments align with the goal of this new business area. Answers b, c, and d are unethical. When meeting with a client to discuss new business opportunities it is best to highlight prior assignments and past accomplishments, and to communicate how these past successes align with your potential client. If you are seek to share money "off the record" or speak in a derogatory nature about other engineering firms you are violating the engineering code of ethics.
3.5	a. A region that exceeds the regulated concentrations for one or more of the seven criteria pollutants. Per the EPA, geographic areas not meeting the national standards for individual criteria pollutants are called "non-attainment areas", therefore the correct answer is a, A region that exceeds the regulated concentrations for one or more of the seven criteria pollutants.
3.6	a. The highest level of a level of a contaminant in drinking water at which no known or anticipated adverse effect on the health of persons would occur. *Per the EPA, the MCLG is the maximum level of a contaminant in drinking water with no known or anticipated adverse effect on the health of individuals, allowing some margin of safety. It should be noted the MCLGs are non-enforceable health goals, which only consider public health and do not take into account the limits of detection or treatment technologies.*

3.7	c. RCRA – Resource Conversation and Recovery Act

RCRA is the federal law that provides the framework for management of hazardous and non-hazardous solid waste. Therefore, all currently generated hazardous waste is regulated by RCRA. |
| 3.8 | a. As the title professional engineer implies, only those with licensing which includes education and experience are qualified to practice engineering that impacts public safety, public welfare, safety, safeguarding of life, health.

The first engineering license law was enacted in Wyoming in the year 1907. All states in the U.S. now regulate engineering practice so that the health, safety and welfare of the public are protected. |
| 3.9 | d. National Pollutant Discharge Elimination System (NPDES)

Per the EPA, the NPDES permit program addresses water pollution by regulating point sources that discharge pollutants to waters of the United States. Therefore, the correct answer is D (NPDES). |

4 – Engineering Economics

4.1	b. 18 years
	There are a few different ways to answer this question. Perhaps the easiest is to utilize the rule of 72, which states that the time to double money is equal to the interest rate times the number of compounding periods (4x18 = 72). Another way to solve is to utilize the Interest rate Tables for i = 4.00%, using F/P, looking for the first "n" greater than 2, which is at n = 18 (F/P of 2.0258).
4.2	b. $13,827
	Use the interest rate tables in the FE Reference Handbook for i = 6%. P = $8,500 + $750 * (P/A, 6%, n = 10) - $350 *(P/F, 6%, n=10) P = $8,500 + $5,520 - $193 P = $13,827.
4.3	c. month 18
	Break-Even is the time to recover $135,000. Each filter provides a net profit to your firm of $14.50 (29.5-15). The number of filters that need to be sold is 135,000/14.5 = 9,310 filters. Each month the firm sells 525 filters. 9,310 filters/525 filters per month = 17.73 months. Therefore, the firm will break-even at 18 months.
4.4	b. $1.10
	The incremental costs are associated with the costs of the additional 20,000 filters. The difference in production costs $22,000 (346,000-324,000) divided by 20,000 (the total number of additional filters) provides the incremental cost per filter of $1.10.

5 – Materials Science

5.1	d. Iron has a higher melt point than aluminum Answer: Utilize the table "Properties of Materials" in the FE Reference Handbook, and you will notice that the melting point of aluminum is 660oC, which would not service temperatures up to 800oC. Therefore, iron is the correct answer.
5.2	d. Epoxy Resin Water, oxygen, and salts (including sodium chloride) will all help to cause or accelerate corrosion (rusting) of iron. Epoxy resins are actually used in certain applications as a coating of iron or steel to prevent corrosion.
5.3	d. Semiconductors Semiconductors are materials which are uniquely able to possess the electrical conductivity between that of a metal and that of an insulator. These are very important materials for computing devices.
5.4	c. Surface Area Activated carbon is widely utilized to capture organic contaminants in water. Among the material properties listed, the very high surface area of activated carbon is what enables effective adsorption of organic compounds. The surface area of activated carbon can exceed 1,000 square meters per gram of material. While other properties such as hydrophobicity, surface functional groups, and pore distribution may contribute to specific performance, surface area is the appropriate selection within the list of potential answers.
5.5	b. 0.00105. Within the FE Reference Handbook, the equations for engineering strain and true strain are listed, with the equation for engineering strain = change in length/initial length. Therefore the answer is 0.0021/2.0 = 0.00105.
5.6	d. 21,000 kg is the correct answer. Utilize the Properties of Materials table in the Handbook and you will find the density of silver is 10,500 kg per cubic meter. Therefore, 2 cubic meters is 21,000 kg.
5.7	a. 0.002 Ohms Utilize the Properties of Metals table to find electrical resistivity of aluminum and use the equation from the FE Reference Handbook in the Material Science section: $R = \rho L/A$ = 0.000000025 Ohm m * 1.25 m / 0.000015 m2 = 0.00208 Ohms
5.8	b. 662 kJ Refer to the section to Materials Science in the FE Reference Handbook and there is an equation for thermal properties and the properties of metals table will provide you the specific heat of iron. = 456.4 J/kg K * (290 K) * (5 kg) = 661,780 J = 662 kJ.
5.9	d. Electromagnetic Induction Answer: One of the keys to this question is to recognize that you need to select which method is NOT a method to control corrosion. Coatings and linings, cathodic protection, and material selection are all common methods to prevent corrosion control. Electromagnetic induction is related to electromotive force.
5.10	a. Yield The correct answer is yield. When a material transitions from elastic to plastic behavior this is called yield.

6 – Environmental Science and Chemistry

6.1	a. 0.014 mg/L Recognize that K' = Cw/Cair Step 1 convert KH to K' (dimensionless), 0.18 * 298 * (1/12.2) = 4.4 = K' Step 2 multiply K' by Cair = Cw (micrograms/m3) = 14,000 Step 3 convert 14,000 micrograms/m3 to mg/L *10-6 = 0.014 mg/L
6.2	d. $C_6H_{12}O_6 + 6O_2 \rightarrow 6CO_2 + 6H_2O$
6.3	b. $FeS + 2 HCl \rightarrow FeCl_2 + H_2S$
6.4	d. $FeCl_3 + 3 H_2O \rightarrow Fe(OH)_3 + 3 HCl$
6.5	c. 9.05 **Answer:** Ksp = $[Cd^{2+}][OH^-]^2$ 4.7 mg/L * (1 g/1000 mg) * (1 mole / 112.4 g) = 0.000041815 M of Cd^{2+} 5.27×10^{-15} = $[OH^-]^2$ [0.000041815] $[OH^-]^2$ = 1.26×10^{-10} $[OH^-]$ = 0.00001122, pOH = 4.95 pH = 9.05
6.6	c. 4.74 By definition, when the acid and conjugate base are equal (0.5 M) the pH is equal to the pKa. Therefore, the correct answer is c. 4.74.
6.7	c. 2.52 pH = -log [H+] = -log (0.003) = 2.52
6.8	d. 5:1 **Answer:** $C_3H_8 + 5 O_2 \rightarrow 3 CO_2 + 4 H_2O$ *Therefore the mole ratio of oxygen to fuel is 5:1.*
6.9	c. 12.60 hours *Answer:* Recognize that 130 mg/L to 65 mg/L is ½ of the initial concentration and therefore you are asked what is the half-life. The half-life for first order reaction is = $0.693/k = 0.693/.055 hr^{-1}$ The answer is therefore 12.60 hours.
6.10	b. 8 ppm/hour To answer this question you need to look at the change in NO_3^- over the 1.5 hours between 1.5 hour and 3 hours. The change is 12 ppm/1.5 hours = 8 ppm/hour.

6.11	**c. Polar and more water soluble** *The carbonyl functional group contains oxygen and imparts uneven electron sharing, which results in a polar functionality. This polar functionality makes acetone more water soluble, as water is itself a polar solvent.*
6.12	**c. meta-xylene** The 1st and 3rd position on the benzene ring identify as meta-
6.13	**a. A process by which metal contaminants are removed from soil via plants.** *Phytoextraction is a sustainable and plant based form of soil remediation. Thefore, the correct answer is a.*
6.14	**c. Nitrogen and Phosphorous** Answer: Eutrophication begins when water environments are enriched with nutrients (N and P)
6.15	**c. Activated Carbon Adsorption and Partitioning from water to organic soils** Kow or octanol water partitioning coefficient provides insight on how organics compounds partition between organic media (carbon containing soil, activated carbon, organic solvents) and water. Therefore, the correct answer is c.
6.16	**b. 25.4 mg/L** **Answer:** *Recognize that the total Mass of compound M = 20 mg which is distributed between two equal volumes of solution. The distribution of $K_{ow} = C_{org}/C_w = 0.05$. $C_{org} = 0.05\ C_w$. Therefore, we can use one equation to solve for C_w* Mass = 20 mg = (C_w * 0.75) + (C_{org} * 0.75) = (C_w * 0.75 L) + (0.05 C_w * 0.75 L) = 0.75 C_w + 0.0375 C_w = 0.7875 C_w C_w = 20 mg/0.7875 L = 25.4 mg/L

6.17	**c. 9,205 µg/L**
	
	Recognize that $K' = C_w/C_{air}$
	   Step 1 multiply K' by C_{air} = C_w (micrograms/m3) = 9,205,000
	   *Step 2 convert 9,025,000 micrograms/m³ to mg/L *10-3 = 9,205 µg/L*

6.18	**b. 150 mg/L**
	
	Recognize that $K' = C_w/C_{air}$
	   Step 1 multiply K' by C_{air} = C_w (micrograms/m3) = 150,430,000
	   *Step 2 convert 150,430,000 micrograms/m³ to mg/L *10-6 = 150 mg/L*

6.19	b.	$2\ Fe + 3\ Cl_2 \rightarrow 2\ FeCl_3$
6.20	c.	$2\ KMnO_4 + 16\ HCl \rightarrow 2\ KCl + 2\ MnCl_2 + 8\ H_2O + 5\ Cl_2$
6.21	c.	4-6 mg/L Answer: It is generally accepted that the minimum level of DO to support fish is 4-6 mg/L.
6.22	c.	6.6 mg/L Answer: DO = DOsat − D = 9.8 − 3.2 = 6.6 mg/L

7 – Risk Assessment

7.1	a. auditory **Answer:** *Inhalation, dermal, and ingestion are three of the routes for toxicology, auditory is not considered a route for toxicology but is a route of noise exposure.*
7.2	b. 4.1x10-4 mg/kg d $$CDI = \frac{\left(0.040\,\frac{mg}{L}\right)\left(2.3\,\frac{L}{d}\right)\left(365\,\frac{d}{yr}\right)(22\,yr)}{(65.4\,kg)\left(365\,\frac{d}{yr}\right)(75\,yr)}$$ CDI = b. 4.1x10^{-4} mg/kg d
7.3	b. 0.011 mg/kg d To start, assume that the exposure duration is to date, so 29 years of exposure. Also, assume that the respiration rate is 15.2 m3/day = 0.633 m3/hour. $$CDI = \frac{\left(0.65\,\frac{mg}{m^3}\right)\left(0.633\,\frac{m^3}{hr}\right)\left(8\,\frac{hrs}{d}\right)\left(250\,\frac{d}{yr}\right)(29\,yr)}{(78\,kg)\left(365\,\frac{d}{yr}\right)(75\,yr)}$$ CDI = 0.011 mg/kg d
7.4	c. 1x10-6 The EPA sets a target cancer risk goal of remediation of 10-6 for soil, groundwater as drinking water, and ambient air.
7.5	b. 1.45x10^{-5} Answer: $Risk = CDI * SF$ $$Risk = 5x10^{-2}\,\frac{mg}{kg\,d} * 2.9\,x10^{-2}\,\frac{kg\,d}{mg}$$ Risk : b. 1.45x10^{-5}
7.6	c. 0.007 mg/kg d $$CDI = \frac{\left(1.0\,\frac{mg}{L}\right)\left(2.3\,\frac{L}{d}\right)\left(365\,\frac{d}{yr}\right)(15\,yr)}{(65.4\,kg)\left(365\,\frac{d}{yr}\right)(75\,yr)}$$ CDI = 0.007 mg/kg d
7.7	d. 0.5 When using a hazard index, the target hazard is less than 1.0. Therefore the correct answer would be d. 0.5.

8 – Fluid Mechanics

8.1	b.	500 cfm
	Answer: *Use Mannings Equation*	
	$$Q = \frac{k}{n} A R_H^{2/3} S^{1/2}$$	
	$Q = \frac{1.486}{0.012} (14.14)(14.14/3\pi)^{2/3} 0.0475^{1/2}$ and therefore Q = 500 cfm	
8.2	b.	228 psi
	Answer: P = γh = 62.4 lb/ft³ * (525 ft) = 32,760 lb/ft² * (1 ft²/144 in²) = 227.5 lb/in²	
8.3	b.	The pump is located at an elevation below the fluid source which flows into the suction side of the pump.
8.4	a.	17 ft
	$$h_L = \frac{4.73\,(1000)}{(150)^{1.852}\,(1)^{4.87}}\, 7.147^{1.852}$$	
	h_L = 16.86 ft	
8.5	a.	0.20 cfs
	Use orifice discharging freely into atmosphere. But first, calculate the initial height of water in the tank, which is 100,000 gallons (0.134 ft³/gallon) = 13,400 ft³, which can then be divided by the area of the tank which is: 13,400 ft² / Pi*(9²) = 52.7 ft, which is the height. $Q = C A_o \sqrt{2gh}$	
	$$Q = 0.65\,(\frac{1}{4}\pi\,(\frac{1}{12})^2)\sqrt{2\left(32.2\frac{ft}{sec^2}\right)(52.7\,ft)}$$	
	Q = 0.20 cfs.	
8.6	c.	39.9 cfs
	$Q = CA\sqrt{2g\,(h_1 - h_2)}$	
	$$= 0.6\,(1.25\,ft^2)\sqrt{2\left(32.2\frac{ft}{sec^2}\right)(107\,ft - 63\,ft)}$$	
	Q = 39.9 cfs.	

8.7	c.	27.6 kN m/sec
		$$W = \frac{1000 \frac{kg}{m^3} \left(9.8 \frac{m}{s^2}\right)(10\ m)\left(0.2 \frac{m^3}{s}\right)}{0.71}$$
		W = 27.6 kN m/sec
8.8	a.	$3300
		This is just a dimensional analysis problem. Recognize that 1 hp-hr = 0.746 kwh, which is stated in the Reference Handbook Table of Conversion Factors near the front of the book.
		365 d/yr * (24 hr/d) = 8,760 hr/yr * 10 hp = 87,600 hp-hr * (0.746 kwh/ 1 hp-hr) = 65,300 kwh 1 kwh costs $.05, therefore 65,350 * $0.05 = $3267 which is ~ $3300.
8.9	d.	43 hp
		First, convert the flowrate to cfs, 1.5 MGD (0.134 ft³/gal) = 201,000 ft³/day
		201,000 ft³/day * (1 d/24 hr) * (1 hr / 60 min) * (1 min/60 s) = 2.326 ft³/sec
		$$W = \frac{62.4 \frac{lb}{ft^3}\left(32.2 \frac{ft}{s^2}\right)(121\ ft)\left(2.326 \frac{ft^3}{s}\right)}{0.74} = 764{,}194 \frac{lb\ ft^2}{sec^3}$$
		1 hp = 17,696 lb ft²/sec³
		W = 764,194/17,696 = 43 hp
8.10	a.	2900 hp
		First, convert the flowrate to cfs, 35 MGD (0.134 ft³/gal) = 4,690,000 ft³/day
		4,690,000 ft³/day * (1 d/24 hr) * (1 hr / 60 min) * (1 min/60 s) = 54.28 ft³/sec
		$$W = \frac{62.4 \frac{lb}{ft^3}\left(32.2 \frac{ft}{s^2}\right)(357\ ft)\left(54.28 \frac{ft^3}{s}\right)}{0.77} = 50{,}565{,}906 \frac{lb\ ft^2}{sec^3}$$
		1 hp = 17,696 lb ft²/sec³
		W = 50,565,906 / 17,696 = 2857 hp
8.11	c.	Use variable speed drives and smaller pumps to improve efficiency
8.12	b.	1.42 cfs
		Q = CH^(5/2) = 2.54 (9.5/12)^(5/2) = 1.42 cfs
8.13	c.	7.54 cfs
		Answer: Q = CLH^(3/2) = 3.33 * (1.75 ft) * (14.25/12)^(3/2) = 7.54 cfs

8.14	d.	1210 gpm
		Answer: Q = 29.8 (D)² (C_d) P^{1/2} = 29.8 * (3)² * (0.90) * (25)^{1/2} = 1207 gpm
8.15	d.	716 feet
		Answer: When pumps are in series the head is additive, H = 367 + 349 = 716 feet.
8.16	c.	1241 ft
		Convert gpm to ft³/sec, 600 gal/min * (0.134 ft³/gal) * (1 min/60 sec) = 1.34 ft³/sec $$h_L = \frac{4.73 \left(5280 \frac{ft}{mi} * 5\ mi\right)}{(100)^{1.852}\ (0.5)^{4.87}}\ 1.34^{1.852}$$ h_L = 1241 feet.

Note: Some of the equations needed to solve problems in fluid mechanics may be found in the Civil Engineering section.

9 – Thermodynamics

9.1	b. 11,364 kJ
	Answer: $q = cm\Delta T = (4.18 \text{ kJ/kg K})*(37.5 \text{ kg})*(72.5 \text{ K}) = 11,364 \text{ kJ}$
9.2	a. 29,400 J
	Use $q = cm\Delta T$
	(170 oC)(0.5 J/g oC) (1000 g) = 85,000 J (170oC) (0.9 J/g oC) (363 g) = 55,646 J 85,000-55,464 = 29,354 J
9.3	c. 18.8 psi
	$P_1/T_1 = P_2/T_2$ 14.7 psi/260K = P_2/333K P2 = 18.8 psi
9.4	d. 141,000 lbs
	This is a problem that can be solved with the ideal gas law, by first finding the volume of the warehouse. $V = 125,000 \text{ ft}^2 * 15 \text{ ft} = 1,875,000 \text{ ft}^3$ (28.2 L/ft^3)
	V = 52,875,000 L PV = nRT 1 atm (52,875,000 L) = n (0.0826 L atm/mol K) * (293 K) n = 2,199,131 moles (29 g /mole) / (1 lb/453.6 grams) = ~ 141,000 lbs
9.5	b. 7,862 J W_b = RT ln(V_2/V_1) = 8.314 J/mol K (288 K) ln (0.75/20) W_b = 7862
9.6	d. 36 g/mole MW = 0.2 (32 g/mole) + 0.35 (28 g/mole) + 0.45 (44 g/mole) = 36 g/mole
9.7	d. 42.8 g/mole MW = 0.1 (32 g/mole) + 0.9 (44 g/mole) = 42.8 g/mole

10 – Water Resources

10.1	b.	1.62 MGD
		2500 * (360 L/d) + 5000 * (285 L/d) + 6250 * (250 L/d) + 11,250 * (200 L/d) = 6,137,500 L/d * (1 gal/3.785 L) = $\boxed{1,621,532}$ gal/d
10.2	c.	5.12%
		Current = 11,200 * 350 L/d * 1 gal/3.785 L = 1,035,667 gal/d
		New Demand = 36 employees * (15 gal/d) + 550 guest * (80 gal/d) + 20 wash. Ma * (425 gal/d) = 540 gal + 44,000 gal + 8,500 gal = 53,040 gal/d
		Percent of Use = 53,040 gal / 1,035,667 gal = $\boxed{5.12\%}$
10.3	a.	Average: 1.71 MGD, Peak: 3.98 MGD
		Residential = 15,000 ppl * (90 gal/capita d) = 1.35 MGD
		Commercial = 1,350 m^3/d * 1000 L/m^3 * (1 gal/3.785 L) = 0.356 MGD
		Average = $\boxed{1.71\text{ MGD}}$
		Peak = Res @ (1.35 MGD * 2.5) + Com @ (0.356 * 1.7) = $\boxed{3.98\text{ MGD}}$
10.4	b.	54,000 gallons
		100 gal/day * 18% = 18 gal/d due to showering
		18 gal/d * 0.2 = 3.6 gal/day for showering
		20,000 ppl * 0.75 utilization * 3.6 gal/day = $\boxed{54,000\text{ gal}}$
10.5	c.	91.7 MGD
		35 MGD * 0.82 = 28.7 MGD * 2.9 = 83.23 MGD
		35 MGD * 0.18 = 6.3 MGD * 1.35 = 8.505 MGD
		$\boxed{91.735\text{ MGD}}$
10.6	b.	watershed
10.7	a.	Surface runoff
10.8	a.	5.7 cfs
		Q = CIA
		Q_1 = 0.33 * 5.1 in/hr * 12.5 acres = 21.04 cfs
		Q_2 = 0.42 * 5.1 in/hr * 12.5 acres = 26.775 cfs
		$Q_{increase} = Q_2 - Q_1 = \boxed{5.74\text{ cfs}}$
10.9	c.	20.3 cfs
		t_c = 50 minutes A_{tot} = 11 + 12 + 24 = 47 acres
		I_{50} @ 50 min = 3.6 inches/hour
		Q = CIA = 0.12 * 3.6 * 47 = 20.3 cfs

10.10	d.	49.9 cfs
		$Q = 0.32\,(3.4)(11) + 0.61\,(3.4)(12) + 0.16\,(3.4)(24)$ $Q = 49.9$ cfs
10.11	a.	34.7 cfs
		$Q = CIA$, $t_c = 22$ mins @ 10 yrs, $I = 3.85$ inches/hr $Q = 0.75 * (3.85\text{ in/hr}) * (12\text{ Acres})$ $Q = 34.65$ cfs
10.12	b.	28.1 cfs
		$t_c = 17$ min @ 50 yr, $I = 5.8$ in/hr $Q = CIA = 0.44 * (5.8) * (11\text{ acres}) = 28.1$ cfs
10.13	b.	62.6 cfs
		$t_c = 50$ min @ 50 yr, $I = 3.6$ in/hr $Q = CIA = (0.12 * 3.6 * 23) + (0.61 * 3.6 * 24)$ $Q = 9.936 + 52.7 = 62.6$ cfs
10.14	a.	13,000 gpm
		$t_c = 17$ min @ 5 yr = 3.3 inches/hour $Q = 0.80 * (3.3) * 11 = 29.04\text{ ft}^3/\text{sec} * (1\text{ gal}/0.134\text{ ft}^3) * (60\text{ sec/min}) = \boxed{13{,}000\text{ gpm}}$
10.15	b.	1.54×10^9 ft³
		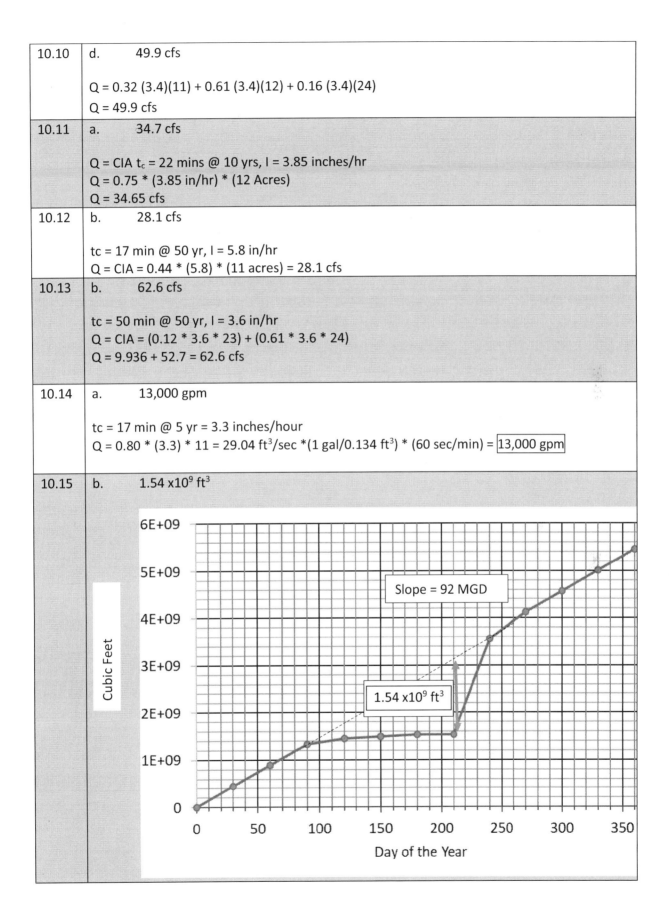

10.16	b.	116,000 gal From hours 1100 to 1700, Sum the difference in Demand – 750 gpm, = (765 – 750) + (821-750) + (949 – 750) + (1007 – 750) + etc… = 115,620 gallons
10.17	c.	The reservoir will likely be undersized
10.18	a.	Sandy and Silty Soil
10.19	d.	Removing nutrients from the soil
10.20	d.	Sediment
10.21	a.	Litter and Improperly disposed refuse
10.22	a.	Non-structural Controls and Structural Controls
10.23	d.	Be sure to disrupt steep slopes and areas with erodible soils

11 – Water and Wastewater

11.1	a.	0.5 mg/L
11.2	c.	18 °C
11.3	c.	35,000 mg/L
11.4	b.	250 mg/L
11.5	d.	40 mg/L as N to 10 mg/L as P
11.6	d.	8870 pounds 2,000,000 gal (3.785 L/gal)(325 mg/L)(1 g/1000 mg)(1 lb/453.6 g) = 5,424 lbs 750,000 gal (3.785 L/gal)(550 mg/L)(1 g/1000 mg)(1 lb/453.6 g) = 3,442 lbs Total = 8,866 pounds.
11.7	b.	6700 gallons 20,000 lbs/0.12 = 166,667 lbs/day – total sludge 166,667 lbs total – 20,000 pounds of dry solids = 146,667 lbs of water/day 20,000 lbs/0.18 = 111,111 lbs/day – total – 20,000 pounds of dry solids 111,111 lbs/day – total – 20,000 pounds of dry solids = 91,111 lbs of water. 146,667 lbs of water (@12%) – 91,111 lbs of water (@ 18%) = 55,556 lbs of water/8.36 lbs/gal = 6650 gal
11.8	a.	2350 pounds per day Use 500,000 gal of GW/day (3.785 L/gal) (0.5 mg/L) = 946,250 mg/d 500,000 gal of SW /day (3.785 L/gal) (22 mg/L) = 41,635,000 mg/d Sum of NOM per day = 42,581,250 mg NOM/day (1 g GAC/ 40 mg NOM) (1 lb/453.6 g) = 2350 lbs/day
11.9	b.	33% **Answer:** 5/15 = 0.33 = 33%

11.10	c.	6,980 sq. ft.
		
		 1.25×10^6 gal/day (0.134 ft^3/gal) = 167,500 ft^3/day
		      V = Q/A
		      A = Q/V = 167,500 ft3/d / 24 ft/d = 6,980 ft^2

11.11	d.	66% Use Table "Typical Primary Clarifier Efficiency Percent Removal" in the FE Reference Book. Then notice that 700 gpd/ft^2 is between 600 and 800 values listed in the column, therefore interpolate. (66 is between 64 and 68).

11.12	d.	Nitrification
11.13	b.	Schmutzdecke
11.14	d.	6.17 m

$$h_f = \frac{1.067 * (0.00243^2) * (1.5\ m) * (48.5)}{\left(9.8\ \frac{m}{s^2}\right)(0.4^4)(297 x 10^{-6}\ m)}$$

= 6.17 m

11.15	b.	0.48

$$n_f = \left(\frac{V_B}{V_t}\right)^{0.22}$$

Vb = 100 L/sec (1 m³/1000L) = 0.1 m³/sec / 9m = 0.0111 m/sec
Vt = 1 ft/sec (0.3048 m/ft) = 0.3048 m/sec

$$n_f = \left(\frac{0.0111\ \frac{m}{s}}{0.3048\ \frac{m}{s}}\right)^{0.22} = 0.48$$

11.16	c.	17.4 pounds

If no units are provided for K and 1/n, it is assumed utilize Ce in terms of mg/L.
qe = $KC_e^{1/n}$ = 120 (0.1 mg/L)$^{0.56}$ = 33.05 mg/g
   7 – 0.1 = 6.9 mg/L – removed
   10,000 gal/d (3.785 L/gal) = 37,850 L/d * (6.9 mg/L) = 261,165 mg Styrene/d
   261,165 mg/d / 33.05 mg/g = 7,902 g-GAC/d (1/453.6 g/lb)
   CUR = 17.4 lbs/day

11.17	d.	trace organic
11.18	b.	400 cf

320 lbs/d * 60 days * 1 ft3/25 lbs = 762 cubic feet

Since there are two vessels, 762 c.f./2 = 384 c.f., round up to 400 c.f.

11.19	a.	55 mg/g

ultimate loading is x/m or qe.

qe = $KC_e^{1/n}$ = 32 (2 mg/L) $^{0.78}$ = 55 mg/g

11.20	d.	PAC causes too much headloss across the vessel, compared to GAC

11.21	b.	4.2
		Rs = 0.234 * 25 = 5.85
		NTU = (5.85/4.85) ln [(1900 µg/L / 50 µg/L) (4.85) + 1]/ 5.85] = 4.17
11.22	d.	7.5 m
		Z = NTU * HTU
		= 1.05 * 7.1 = 7.455 m
11.23	a.	11
		H' = H/RT
		$$H' = \frac{6.4 \times 10^{-3} \frac{atm\, m^3}{mol}}{8.2057 \times 10^{-5} \left(\frac{atm\, m^3}{mol\, K}\right)(291\, K)} =$$
11.24	d.	The air stripper performance would be more effective
11.25	b.	The air compressor, because this would increase the stripping factor
		If you plug in different Rs values, from Rs = H' (Qa/Qw), say 4 and 40 into the NTU equation, it becomes evident that in an established air stripper, a computed lower NTU value equates to more effective air stripping.
11.26	a.	6.8 grams
		50 tons * 0.3 = 15 tons – solids/yr 15 tons – solids * (2000 lbs/ton) * (1 kg/2.2 lbs) * (0.5 mg Zn/ kg solid) * (1 g/1000 mg) = 6.8 g Zn/ yr for each acre.
11.27	c.	$60,000

		23 tons/d * $30/ton * 365 d/yr = $251,850/yr
		    23 tons * 0.16 = 3.68 tons solids/d
		    3.68 /0.21 = 17.5 tons/d
		    17.5 tons/day * ($30/ton) * ($365 d/yr) = $191,885/yr

		$251,850 – 191,885 = $59,964/yr
11.28	c.	Option 3
		Option 1: 5.5 MGD * 0.2 * 0.3 = 330,000 gal Option 2: 5.5 MGD * (8/124.9) * (0.15) = 52,842 gal Option 3: 5.5 MGD * (28/124.9) * (0.65) * (0.5) = 400,720 gal Option 4: 5.5 MGD * (0.5) * (0.3) * (35/124.9) = 231,185 gal

11.29	d.	Utilizing a flat billing rate structure for users, where all users pay the same for water regardless of use.
11.30	d.	91 s^{-1} $$G = (P/mV)^{0.5}$$ $$G = (50 \text{ W}/(0.798 \times 10^{-3} \text{ N s/m}^2 * 7.5 \text{ m}^3))^{0.5}$$ $$G = 91.4 \text{ s}^{-1}$$ Therefore, d is the correct answer.
11.31	d.	120 Individual divalent cation Hardness (mg/L of $CaCO_3$) = M^{2+} (mg/L) x (50/EW of M^{2+}) The total hardness of water is the sum of the individual divalent cation hardness. 	Cation	EW	Hardness mg/L of $CaCO_3$	|---|---|---|	Ca^{2+}	20	38*50/20 = **95**		Mg^{2+}	12.2	6.1*50/12.2 = **25**	 The Hardness is equal to 120 mg/L of $CaCO_3$, therefore the answer is d.
11.32	b.	80 mg/L Use the lime-soda softening equations, and since the calcium carbonate hardness is provided ($CaCO_3$), the mole ratio to slaked lime ($Ca(OH)_2$) is equivalent, or a 1:1 ratio. Therefore, we need to do mole to mass conversions. (105 mg/L of $CaCO_3$)*(100 g / mole of $CaCO_3$)*(1 mole of $Ca(OH)_2$ / 74 g) = 77.8 mg/L* (1/0.95) = 77.8 mg/L * (1/0.96) = 80 mg/L of slaked lime
11.33	b.	7 days Since the system is completely mixed, we can assume that $X_A = X_W = X_E$ SRT = $V(X_A) / (Q_W X_W + Q_E X_E)$ SRT = 350,000 L * 2,950 mg/L / (12,500 L/d * 2850 mg/L) + (37,500 L/d * 2850 mg/L) SRT = 7 days, therefore the answer is b.
11.34	a.	6600 L/min g 1.5 x10^6 gallons/day * 3.785 L/gal * 1 day/24 hours * 1 hour/60 minutes = 3942.7 L/minute Solids Loading Rate = QX_A = 3942.7 L/minute * (1,675 mg/L) * (1 g/1000 mg) Solids Loading Rate = 6604 L/min g

11.35	d.	1220 mg/L
		HRT = O = 2300 m³/575 m³/hr = 4 hours
		$X_A = O_c Y (S_0 - S_e)/O(1 + k_d O_c)$
		$X_A =$ 8.2 days (0.9 kg biomass/kg BOD) * (0.31 kg BOD/m³ - 0.88*0.31 kg BOD/m³)
		4 hours/(24 hrs/day) * (1 + 0.05 days^{-1} * 8.2 days)
		$X_A =$ (0.2745 days * kg biomass/m³) / 0.2256 days
		$X_A =$ 1.2167 kg biomass/m³ (1000 g/kg) (1000 mg/g) (1 m³/1000 L)
		$X_A =$ 1216.7 mg biomass/L
11.36	c.	Install the decentralized treatment system to serve the community.
11.37	b.	183 mg/L
		The hydraulic loading rate is is 100 ft³/min divided by 75 ft², which is 1.33 ft³/ft² min, which is also 1.33 m³/m² min.
		$S_e/S_o = e$^$-(kD/q^n)$
		S_e = (295 mg/L) * e^$-(0.06$ min^{-1}*9.144 m / 1.33^0.5)
		S_e = 183 mg/L
11.38	c.	3 – log
		T10 = HRT * BF = 15 minutes * 0.5 = 7.5 minutes
		CT = C x t10 = 0.35 mg/L * 7.5 minutes = 2.65 = 3-log removal of bacteria, answer is C.
11.39	d.	9 minutes
		Requires CT value of 3. (P 204)
		Need 4-log removal (p 203)
		CT = 3 mg min/L = 0.5 mg/L * t10
		. t10 = 6 min
		. t10 = HRT * BF
		HRT = 6 min/0.7 = 8.6 minutes = 9 minutes, answer d
11.40	d.	Course screens, fine screens, grit removal, equalization, primary treatment
11.41	a.	Primary Settling, Biological Processes, Secondary Settling, Filtration, Disinfection
11.42	a.	4275 m3
		$$\frac{20}{590} = \frac{1}{1+0.1\tau} = 0.33898$$
		$$0.33898 = \frac{1}{x}$$
		$$29.5 = 1 + 0.1 t$$
		$$\tau = 285 \; days$$
		Since V/Q = 285 days, Q = 15,000 L/day = 4275 m³

11.43	d.	11 days
11.44	c.	50.1 kmol/m2 sec J_w = 3.1x10-5 kmol/m^2 s Pa * (1620 kPa) = 50.2 kmol/m^2 sec
11.45	a.	2225 lbs of oxygen/day 1 mole of NH_4^+ requires 2 moles of O_2, which means 18 pounds of NH_4^+ requires 64 pounds of Oxygen (18 lbs NH4+/64 lbs O) 2.5 MGD * 3.785 L/gal * (30 mg NH_4^+/L) * (1 g/1000 mg) * (1 lb/453.6g) * (64 lb O/18 lb NH_4^+) Pounds of oxygen required = 2225 lbs/day
11.46	b.	The introduction of 0.5 mg/L of chromium and 2.0 mg/L of copper. *In fact, a shift in operating pH from 7.0 to 7.5 would move nitrification towards optimum pH range. The increase in temperature would not inhibit nitrification and the increase in dissolved oxygen would not inhibit – as this is an oxidation process. Therefore, nitrifying bacteria are very sensitive to chemical upsets, particularly metals. Although trace metals have been found to be helpful, these concentrations would likely be inhibitory. Therefore, the answer is b.*
11.47	c.	H+ < Na+ < K+ < Zn2+ The common selectivity of cations would indicate that c. H+ < Na+ < K+ < Zn2+ is the correct series order for the cations listed.
11.48	d.	43 L 30 mg/L / 65.4 g/mole / 2 meq/mole = 0.8 meq/L * 12000 L = 9600 meq 960 meq / (375 meq/kg) = 25.6 kg 25.6 kg / 0.6 kg/L = 43 L
11.49	d.	0.20 m $$V_z = 1.5\,m * \frac{20{,}450 - 17{,}835}{20{,}450 - 0.5\,(2615)} = 0.20\,m$$
11.50	b.	371 g Answer: x/m = aKCe / (1 + KCe) = (155 * 0.05 * 5) / (1 + (0.05 * 5)) = 31 mg/g 120 mg/L – 0.05 mg/L = 115 mg removed/L * 100 L = 11,500 mg / 31 mg/g = 371 g

12 - Air Quality

12.1	b.	9985 ft²
		$n = 1 - e^{(-wA/Q)}$
		Q = 100,000 cfm (1 min/60 sec) = 1667 cfs
		A = 9987 ft²
12.2	c.	0.085 g SO2/g Fuel
		$C_{55}H_{62}S + 71.5\ O_2 \rightarrow 55\ CO_2 + 31\ H_2O + SO_2$
		SO2 = 64 g/mole, therefore EI = 64/755 = 0.085
12.3	a.	69.3%
		$VOC_{reduction}$ = 70% * 99% = 69.3%
12.4	b.	97.80%
		DRE = (Win – Wout)/Win * 100%
		DRE = (91-2)/91 * 100% = 97.8%
12.5	d.	Cyclone for Scenario 1 and Baghouse for Scenario 2.
12.6	a.	A primary pollutant is emitted directly from a source whereas a secondary pollutant is formed via atmospheric reactions.
12.7	c.	44,444 m2
		From Table provided: 0.9 m³/min m²
		A_{min} = 40,000 cfm/0.9 m³/min m² = 44,444 ft²
12.8	b.	62 mg/m³
		$$Cmax = \frac{Q}{\pi u\ \sigma_y \sigma_z} e^{\left(\frac{-1}{2} \frac{H^2}{\sigma_z^2}\right)}$$
		$$:Cmax = \frac{5{,}087{,}000\ mg/sec}{\pi \left(6 \frac{m}{sec}\right)(75\ m)(33 m)} e^{\left(\frac{-1}{2} \frac{35^2}{33^{\wedge}2}\right)}$$
		Cmax = 62 mg/m³

12.9	b.	20 ft³
	100 lbs of benzene (100 lbs C/17 lbs Benzene) = 588 lbs C (1 ft³/30 lbs) = 19.6 ft³	
12.10	D.	345 µg/m3

$$Cmax = \frac{300{,}000{,}000 \; \mu g/sec}{\pi \left(9 \dfrac{m}{sec}\right)(220\,m)(130\,m)} \; e^{\left(\frac{-1}{2} \frac{50^2}{130^{\wedge}2}\right)}$$

Cmax = 345 micrograms/m³

12.11	c.	0.095
	Answer: CH4 + 2 (O2 + 3.76 N2) → CO2 + 2H2O + 7.52 N2, YCO2 = 1 / (1 + 2 + 7.52) = 0.095	
12.12	c.	0.155
	C_3H_8 + 5 (O_2 + 3.76 N_2) → 3CO_2 + 4H_2O + 18.8 N_2, Y_{h2o} = 4 / (3 + 4 + 18.8) = 0.155	
12.13	d.	Thermal NOx is formed at high combustion temperatures and fuel NOx is from combustion with nitrogen bound in the fuel.
12.14	a.	170 lbs
	C	= 400,000 ppb * (1 atm * 78 g/mole)/(0.0821 L atm/mol K * 298.15 K) = 1,274,608 µg/m³ * (20 m³/min)(60 min/sec)(8 hrs) (1 mg/1000 µg) = 12,236 g Benzene (100 g GAC/ 16 g B) = 76,476 g GAC * (1lb/453.6 g) = 169 lbs
12.15	d.	Non-Polar Volatile Organics
	You probably know that it is organics, therefore you have a 50/50 chance on this question knowing that. Non-polar VOCs typically adsorb better than polar compounds.	
12.16	a.	15.15 ft³
	50 lbs of E-B (100 lbs of GAC/ 11 lbs E-B) = 454.5 lbs GAC * (1 ft³ / 30 lbs) 15.15 ft³	

12.17	c. $C_4H_{10} + 6.5\ O_2 \rightarrow 4\ CO_2 + 5\ H_2O$	
12.18	d.	Isopropanol
12.19	d.	Activated Carbon
12.20	d.	1244 mg/m³ 525 * (58 / 0.0821 * 298.15) = 1244 mg/m³ (See equation in FE Reference Handbook)
12.21	b.	2310 942 * (60 / 0.0821 * 298.15) = 2310 μg/m³
12.22	d. D See the Atmospheric stability under various conditions chart in the FE reference handbook	
12.23	b.	A common radioactive noble gas
12.24	d.	Radon is the second leading cause of lung cancer
12.25	a.	2880 g 0.2 g/mi * 1,600 sedans * 9 mi/sedan = 2,880 g
12.26	c.	NOx and SO2
12.27	d.	96.4% CE = 85,000 / (85,000 + 3,200) = 96.37%
12.28	b.	33% n = 1 / (1 + 149/74) = 33%
12.29	c.	5.50 Ne = 1/0.5 [1.75 + 2.0/2] = 5.5
12.30	a.	5.00 m See the chart provided in the FE Reference Handbook that provides the cyclone ratio of dimensions to body diameter. Notice that the listed inlet width, W for a high efficiency cyclone is 0.21. Therefore, double the value for the cone length of 2.50 and you will have an answer of 5.0 m.

13 – Solid Waste

13.1	c.	9700 BTU/Lb HHV (BTU/lb) = 145.7 (C) + 619 (H) + 45.1 (S) – 59.8 (O) HHV (BTU/lb) = 145.7 (40) + 619 (8) + 45.1 (5) – 59.8 (22) = 9690 BTU/LB
13.2	c.	Portland Cement
13.3	b.	1000 m^3 15,000 p * (2 lbs/ p day) * (7 d/wk) * (1 kg/2.2 lbs) * (1 m^3/95 kg) = 1005 m^3
13.4	d.	Distance from a military base
13.5	a.	Resource Conversation and Recovery Act
13.6	a.	TCLP
13.7	c.	99.44% DRE = (Win – Wout)/Wout * 100% = (0.21 – 0.001175)/(0.21) * 100% = 99.44% Win = 63 lbs/hr * 0.2 = 12.6 lbs – POHC/hr * (1 hr/60 min) = 0.21 lbs-POHC/min Wout = 4.7 lbs/min * 0.025/100 = 0.001175 lbs POHC/min
13.8	c.	228,125 25,000 lbs/day * 365 d/yr * 30 yr * (1 yd^3/1200 lbs) = 228,125 yd^3
13.9	c.	Hazardous Waste Secure Landfill
13.10	b.	131 days
13.11	a.	3.07x10^{-7} g/cm^2 second Na = (0.05 cm^2/sec) (0.2)$^{4/3}$ (2.45x10^{-4} g/cm^3 – 350x10^{-4} g/cm^3) / (20 cm) Na = 3.07x10^{-7} g/cm^2 sec Note the unit conversion required, 1000 mg = 1 g and 1,000,000 cm^3 = 1 m^3
13.12	d.	A waste mixture that has a pH of 1.25, but contains 50% water
13.13	d.	When shipping hazardous waste materials to disposal
13.14	b.	1910 pounds SWp = SWi + P/(a + bp) = 1080 lb/yd^3 + 400 psi / 0.16 (yd^3/in) + 0.0008 yd^3/in (400 psi) **SWp = 1913 lbs**

13.15	a.	Clay possesses a very low coefficient of permeability
13.16	d.	50% Methane, 45% Carbon Dioxide, 5% Other Gases
13.17	d.	Less economic security for the municipality budget
13.18	d.	Perform an on-site recycling program
13.19	c.	Individual Fines and jail time
13.20	d.	Recycled aluminum needs 90% less energy than new aluminum from bauxite ore
13.21	d.	291.16 1000 gallons (0.134 ft³/gal)(62.4 lbs/ft³)(0.925)(0.083)(1kg/2.2 lbs)(0.9999) = 291.16 kgs. This can very slightly based on your rounding if you performed the calculations step-wise.
13.22	d.	Potentially Responsible Party
13.23	c.	1.6×10^6 BTU 500 lbs (2000 BTU/lb) = 1,000,000 BTU 83 lbs * 7200 BTU/lb = 597,600 BTU Sum = 1,597,600 BTU
13.24	b.	10.9×10^6 BTU 800 lbs * 2,000 BTU/lb 480 lbs * 7,200 BTU/lb 120 lbs * 7,000 BTU/lb 300 lbs * 14,000 BTU/lb 100 lbs * 8,000 BTU/lb 140 lbs * 60 BTU/lb 60 lbs * 300 BTU/lb Sum = 10,922,400 BTUs/ton
13.25	b.	Sodium Sulfide <u>Sodium Hydroxide</u> $5.27 \times 10^{-15} = [x][2x]^2$ $= 4x^3$ $x^3 = 1.3175 \times 10^{-15}$ $x = 1.096 \times 10^{-5}$ M 1.096×10^{-5} M (112.4 g/mole) (1000 mg/g) **= 1.23 mg/L Cd²⁺** <u>Sodium Carbonate</u> $6.18 \times 10^{-12} = [x][x]$ $x = 2.486 \times 10^{-6}$ M 2.486×10^{-6} M (112.4/mole)(1000 mg/g) **= 0.279 mg/L Cd²⁺** <u>Sodium Sulfide</u> $1.40 \times 10^{-29} = [x][x]$ $x = 3.74 \times 10^{-15}$ M 3.74×10^{-15} M (112.4 g/mole)(1000 mg/g) **= 4.21 × 10⁻¹⁰ mg/L Cd²⁺**

13.26	c.	Were there many disgruntled workers that performed illegal activities?
13.27	c.	This is not acceptable; the compounds will generate heat and toxic gases.
13.28	d.	This is not acceptable; the compounds will generate heat, fire, and toxic gases.
13.29	a.	This is fine, there is no reaction expected.
13.30	a.	This is a list of worst sites identified for long-term cleanup.
13.31	b.	To assess the nature and extent of contamination, and understand treatability
13.32	a.	Precipitate and concentrate the radioactive metals.
13.33	a.	High level wastes
13.34	d.	To utilize deep underground isolation of dangerous radioactive wastes.

14 – Groundwater and Soils

14.1	b.	Benzene
	Benzene has a lower MW, higher solubility, and a lower log Kow – which indicate more favorable properties for transport than lindane.	
14.2	d.	Phytoremediation
	Since the contamination is only in the first 8 inches of soil, these leaves options c and d. Some favorable background knowledge would allow you to know that certain ferns are favorable for arsenic removal, leaving Phytoremediation as the obvious solution.	
14.3	a.	Soil Vapor Extraction
	There are few favorable context clues, first the contaminant is light-weight volatile hydrocarbons (suggest stripping) and the gw table is 20 feet below surface and the contamination is above the gw table, confirming that SVE would be best suited here.	
14.4	c.	Bioremediation
	Bioremediation is the best option since it is a groundwater that is contaminated with BTEX.	
14.5	b.	32.1 cfs
	$$Q = \frac{\pi K \left(h_2^2 - h_1^2\right)}{\ln\left(\frac{r_2}{r_1}\right)} \qquad : Q = \frac{\pi \left(0.007 \frac{ft}{sec}\right)\left(104^2 - 67^2\right)}{\ln\left(\frac{38}{0.5}\right)}$$ $$Q = 32.1 \text{ cfs}$$	
14.6	d.	80%
	Porosity, S = Vw/Vv * 100% = 8 cm³/10 cm³ * 100% = 80%	
14.7	d.	8.4×10^{-8} cm/sec
	K = q/iA q = 0.45 cm³/hr * (1 hr/60 min) * (1 min/60 sec) = 1.25×10^{-4} cm³/sec A = πr² = 78.54 cm² K = (1.25×10^{-4} cm³/sec) / (19 * 78.54 cm²) = 8.4×10^{-8} cm/sec	
14.8	d.	Clay
	Clay (<0.000001 cm/sec) vs fine sand (0.001 cm/sec), course sand (0.01 cm/sec) and clean gravel 1.0 (cm/sec)	

14.9	a.	10 feet
	The correct answer is 10', since the zone of intermittent saturation is the variation between the highest and lowest level of the water table.	
14.10	c.	0.00085
	First calculate the difference between GW levels. Point A GW is 190 feet above MSL, Point B GW is 183 feet above MSL. The difference in height is 190-183 = 7 feet.	
	Second, convert 1.55 miles to feet 1 mile = 5,280 feet, 1.55 * 5280 = 8184 feet.	
	Hydraulic gradient = i = dh/L = 7 feet / 8184 feet = 0.000855	
14.11	c.	Soil Permeability
14.12	b.	A well where water flows above the ground surface due to confined water pressure.
14.13	d.	36.6 m/yr
	v = q/n = -K dh/n dx = -0.0058 cm/sec / 0.175 * (-.0035) = 1.16×10^{-4} cm/sec	
	1.16×10^{-4} cm/sec * (60 sec/min) * (60 min/hr) * (24 hr/day) * (365 d/yr) = 3658 cm/yr	
	3658 cm/yr * 1 m/100 cm = 36.58 m/yr	
14.14	c.	7485 days
	750 m * 100 = 75,000 cm / 10.02 cm/day = 7485 days	
14.15	c.	2.52 kg/L
	water content = (Ww/Ws) * 100 = 0.072 = Ww/Ws = Ww/ (1.35 kg – Ww)	
	0.072 = Ww / (1.35 – Ww)	
	Ww = 0.0907 kg	
	Dry unit weight = Ws/V = 1.35 kg – 0.0907 kg / 0.5 L = 2.52 kg/L	